Biochemical characteraztion of (CA15-3) in Sera and Tissues of Breast Tumors I

Prof.Dr.Sami AlMudhaffar and
Dr .Salwa Hameed Nasir1. Al-Rubae'i

This book

deals with Determination of carbohydrate antigen 15-3 (CA15-3) and carbohydrate antigen 19-9 (CA 19-9) levels in sera of patients with benign and malignant breast tumors , Development of IRMA assay for the determination of CA15-3 and CA19-9 from cytosolic tissues of benign and malignant breast tumors ,Characterization of the binding of 125I-anti CA15-3 antibody with isolated human − CA15-3 in benign and malignant breast tumors, such as those of binding capacity and the effect of various factors (pH, temperature, time, halides, salts, CA15-3 and its antibody concentration) ,Determination of kinetic and thermodynamic parameters of the binding of partially purified CA15-3 with its specific antibody and Spectroscopic studies on (125I-anti CA15-3 antibody / CA15-3) complex in breast tissue .

Further more the level of sera CA 15-3 in patients with benign and malignant breast tumors (preoperative) was measured by Immunoradiometric Assay (IRMA). Data analysis showed, concentrations of CA 15-3 were significantly higher in pre-and post-menopausal malignant breast tumors (P< 0.0001) and significantly lower in benign breast tumors, compared with healthy subjects.

.

The results obtained revealed higher incidence of CA 15-3 in two groups of malignant breast tumors than those in benign breast tissues and in the supernatant fraction more than the pellet fraction .The binding of ^{125}I –anti CA 15-3 antibodies with CA 15-3 was studied in three groups: benign breast tumor (Fibroadenoma), pre- and post- menopausal malignant breast tumors (IDC). The optimum conditions observed for the binding were as follows:

CA 15-3 in concentration tissue homogenate 100µg. mL^{-1} for groups II and I while it was 200 µg.mL^{-1} for group III. ^{125}I -anti CA 15-3 antibody concentrations: 0.175µg.mL^{-1} for group I and II, whereas it was 0.140 µg.mL^{-1} for group III. Temperatures of incubation were: 45°C for groups I and III, 15 °C for group II, while time of incubation was 90 min for both group I and III, 30 min. for group II. The optimum pH was 7.0 for group I, 7.6 for group II and 7.8 for group III. The use of different halides was shown to increase the binding between CA 15-3 and ^{125}I –anti CA 15-3 antibody in both group II and III, while inhibition occurred on the binding in group I.

.

Chapter one

Introduction

The breast is constantly responding to changes in hormonal, nutritional, genetic, psychological, and environmental stimuli such as radiation that cause continual cellular changes [1]. As a result of these changes, breast tumors (abnormal breast tissue) may develop either benign (noncancerous) or malignant (cancerous) [2]. The major significance of the benign processes less in the need to separate them from malignancies. The World Health Organization (WHO) classifies tumors of the breast (1981) according to histological aspects [3] (Table 1.1).

Table (1.1): Histological classification of breast tumors [3]

I. Epithelial Tumors
A. Benign
1. Intraductal papilloma
2. Adenoma of the nipple
3. Adenoma
a. *Tublar*
b. *Lactating*
B. Malignant
1. Non invasive
a. *Intraductal carcinoma*
b. *Lobular carcinoma*
2. Invasive
a. *Invasive ductal carcinoma*
b. *Invasive lobular carcinoma*
3. Paget's disease of the nipples
II. Mixed Connective Tissue and Epithelial Tumors
a. *Fibroadenoma*

Table (1.1): Continued.

b.	*Phyllodes tumor (cystosarcoma phyllodes)*
c.	*Carcinosarcoma*
III. Miscellaneous Tumors	
a.	*Soft tissue tumors*
b.	*Skin tumors*
c.	*Tumors of haemopcietia and lymph tissues*
IV. Unclassified Tumors	
V. Mammary Dysplasia / Fibrocystic Change	
VI. Tumor like Lesion	
a.	*Duct ectasia*
b.	*Inflammatory pseudotumors*

1.2. Benign Breast Tumors

1.2.1. Fibroadenoma

This is the most common benign tumor of the female breast. It is a new growth composed of both fibrous and glandular tissue [4]. These tumors are commonly found in younger women between the ages 20-35 years [5]. It increases in size, during pregnancy [6]. It is less likely to develop after menopause. An epidemiological study suggests that fibroadenoma represents a long-term risk for breast carcinomas and that risk is increased in women with ductal hyperplasias, or a family history of breast carcinoma [7].

1.2.2. Fibrocystic disease

This is an ill-defined condition of the breast where palpable lumps can be felt and is usually associated with pain and tenderness that fluctuates with the menstrual cycle [8]. Fibrocystic changes are the most common, occurring in approximately 60% of premenopausal women. Women with this disease usually have a freely movable, palpable mass, at time it may cause pain, particularly when women are in the premenopausal phase of the menstrual cycle, however, breast pain can be caused by lesions other than fibrocystic changes. The palpable lesion may appear to increase and decrease is size cyclically; usually achieving its maximum size in the premenstrual phase of the menstrual cycle. Cystic disease is frequently accompanied by varying degrees of epithelial hyperplasia in adjacent ducts and lobules [9]. In patients who have one particular form of fibrocystic disease (the proliferate form), the incidence of cancer increased very slightly [10].

1.3. Malignant Breast Tumors

1.3.1. Incidence

Breast cancer is the most common malignant tumors in women [11], and it is the leading malignancy affecting women in North America and Europe. In 2000, approximately 184200 new cases of invasive breast cancer were diagnosed in the United States. The number of noninvasive breast cancer is hard to verify, but it probably account for and an additional 20000 to 30000 new cases; thus, the number of invasive and noninvasive breast cancer treated in 2000 approximately 200000 [12].

In Iraq, according to the results of Iraqi Cancer Registry (ICR), breast cancer accounting for (31.11%) and remained the commonest tumors in the year 2000 [13-15], it was also shown that breast cancer was the first among the commonest ten cancers in Iraq. Figure (1.1) represents the population of breast cancer in Iraq for the last nine years. As shown in the same figure, the population of breast cancer increased more than double in the last nine years [16].

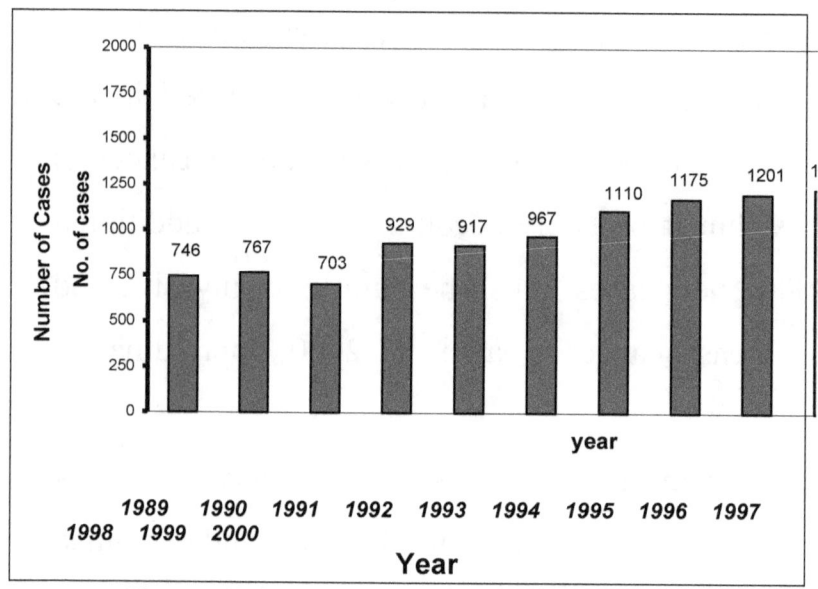

Figure (1.1): The population of breast cancer in Iraq through
(1989-2000)[13-16].

1.3.2. Etiology and Risk Factors in Breast Cancer

Numerous risk factors have been associated with the development of breast cancer, such as genetic, environmental, hormonal, and nutritional.

Despite all available data on breast cancer risk factors, 75% of women with this cancer have not exposed to any risk factors [17].

1.3.2.1. Genetic Factors

Breast cancer is the result of mutations in one or more critical genes. Two genes in women on chromosome 17 have been implicated. The most important gene is called BRCA-1; the other is the P_{53} gene. A third gene is BRCA-2 on chromosome 13 [11].

1.3.2.2. Increasing Ages

Breast cancer is uncommon before age 25 years, but then there is a steady rise to the time of menopause, followed by a slower rise throughout life. The average age at the diagnosis is 64 years [4].

1.3.2.3. Family History

The overall relative risk of breast cancer in women with a positive family history in a first –degree relative (mother, daughter, or sister) is 1.7. Premenopausal onset of the disease in a first–degree relative is associated with three-fold increase in breast cancer risk, whereas postmenopausal diagnosis increases relative risk by only 1.5. When the first degree has relative bilateral disease, there is fivefold increase in risk. The relative risk for a woman whose first-degree relative developed both bilateral and premenopausal breast cancer is nearly nine. No increased risk has been demonstrated when only a second-degree relative (aunt, cousin, or grandmother) has had breast cancer [17].

1.3.2.4.Proliferative Breast Disease

The diagnosis of certain condition after breast biopsy is also associated with an increased risk for the subsequent development of invasive breast cancer[17] . Women with proliferative disease of the breast with a typical hyperplasia (atypia) are at increased risk for developing breast cancer (five-fold increase), however. The risk for atypia is greater in patient with a strong family history of breast cancer (11-fold increase) [11] .

1.3.2.5. Personal Cancer History

A personal history of breast cancer is significant risk factor for the subsequent development of a second, new breast cancer. This risk has been estimated to be as high as 1% per year from the time of diagnosis of the initial cancer. Women with a history of endometerial, ovarian, or colon cancer also have a higher likelihood of developing breast cancer than those with no history of these malignancies [17].

1.3.2.6. Menstrual and Reproductive Factors

Early onset of menarche (<12 years old) has been associated with a modest increase in breast cancer risk (two fold or less). Women who undergo menopause before age 30 have a twofold reduction in breast cancer risk when compared to women who undergo menopause after age 55. A first full-term pregnancy before age 30 appears to have a protective effect against breast cancer, whereas a late first full-term pregnancy or nulliparity may be associated with higher risk, There is also suggestion that lactation protects against breast cancer development [17].

1.3.2.7. Environmental Factors

Women exposed to therapeutic radiation or after atom bomb exposure have a higher rate of breast cancer. Risk increases with younger age and higher radiation doses [4]. Radiation is believed to cause 1-2% of all cancer deaths[18,19]. In respect to pollution studies, which have shown that there is a well-established correlation between many pollutants and cancer, it has been estimated, that 1% of cancer deaths is due to air, water and land pollution [20].

1.3.2.8. Hormonal Influences

Endogenous estrogen excess, or more accurately, hormonal imbalance, clearly plays a significant role. Many risk factors mentioned-long duration of reproductive life, nulliparity, and late age at first child-imply increased exposure to estrogen peaks during the menstrual cycle [4].

1.3.2.9. Dietary Factors

Diets that are high in fat have been associated with an increased risk factor for breast cancer [21]. It has been suggested that differences in dietary fat content may account for the variations in breast cancer incidence observed among different countries. Sala et.al, illustrated that certain macronuterients and food such as protein, carbohydrate and meat intake influence of risk of breast cancer through their effects on breast tissue morphology [22]. Data from prospective studies have confirmed that the relationship exists between alcohol intake and risk of developing breast cancer [23].

Alterations in endrogenous estrogen levels secondary to obesity may enhance breast cancer risk [17].

1.3.2.10. Lactation

In the search of practical methods to prevent breast cancer, lactation has strong evidence as a potentially modifying factor especially at early age and for long period. There is a significant reduction in the risk of breast cancer associated with lactation for more than two years. This effect appeared to be limited to premenopausal women [24]. Lactation may reduce the risk of breast cancer by interrupting ovulation or by modifying pituitary and ovarian hormone secretions. Direct physical changes in the breast that accompany milk production may also contribute to prevent the effect [25].

1.3.3. Histopathology of Breast Cancer

Most cancer of the breast is a carcinoma of the epithelial cells that line breast ducts and lobules. Rarer forms of cancer occurring in the breast arise from the stromal cells that surround the epithelial glands [12]. Breast cancer is a complex, devastating diseases and the most frequently diagnosed cancer in women. It is the single leading cause death for women of age 20-59 years [26].

There are a various type of breast carcinomas according to (WHO) classification [3] :

1.3.3.1. In Situ Carcinoma (Non-Spreading Type)

A. Ductal Carcinoma In Situ (DCIS)

The malignant cells in this disease are confined to the ductal basement membrane [27]. DCIS usually occurs without forming a mass because there is no scirrhous component [11]. DCIS is also known as intraductal carcinoma.

B. Lobular Carcinoma In Situ (LCIS)

LCIS is composed of smaller lobular or acinar cells and fills the terminal breast lobule with a homogenous proliferation; most clinicians currently regard LCIS as a risk factor for the development of invasive breast cancer. LCIS is usually not treated, but affected women are placed under frequent   surveillance [28,29].

1.3.3.2. Infiltrating Carcinoma (Invasive Carcinoma)

A. Infiltrating Ductal Carcinoma (IDC)

Most invasive carcinoma of the breast is ductal in origin [11]. Infiltrating (invasive) breast carcinoma differs from intraductal carcinoma (ductal carcinoma in situ) by the presence of stromal invasion, through which tumor cells spread not only locally but also regionally and distantly via vascular lymphatic space [30].

B. Infiltrating Lobular Carcinoma (ILC)

This type is rare, from about (5%-10%) of breast cancer. It is begins in the milk-secreting glands of the breast. It is often multicentirc, several areas of thickening may occur in one or both breasts. It is characterized by the presence of small and relatively uniform tumors cell growing singly around lobules involved by in situ lobular neoplasia [31].

1.3.4. Staging System of Breast Cancer (11)

The standard staging system for breast cancer is the TNM system table (1.2). The TNM classification devised by the International Union Against Cancer (UICC) and accepted by the American Joint Commission on Cancer Staging a world standard [32]. Another system is the Colombia Clinical Classification (CCC), formulated by Haagensen and Stout [33]. Although this system was a valuable precursor and is easier to remember than the TNM, it is a less precise classification where stage A represents a tumor confined to the breast; stage B include tumors with clinical axillary lymph node enlargement; stage C represents the presence of grave prognostic sings in the breast; and stage D indicates metastatic disease. The TNM based on the clinical features of tumor (T), the regional lymph nodes (N), and the presence or absences of distant metastases (M) [34]. The purposes of staging are the following [23]:

- Plane a therapeutic strategy that most appropriate for the patient.
- Allow for more intelligent prognostication of the disease statues of the patient.

- Permit comparison of therapeutic results obtained from different sources by different means.

Table (1.2): TNM System and Stage Grouping [35].

T	Primary tumor
T0	No evidence of primary tumor
Tis	Carcinoma in situ
T1	Tumor 2 cm or less in greatest diameter
T2	Tumor more than 2 cm but less than 5 cm in greatest diameter
T3	Tumor more than 5 cm in greatest diameter
T4	Tumor of any size with direct extension to chest wall or skin
N	Regional lymph nodes

Table (1.2): Continued.

N0	No regional lymph node metastases
N1	Metastasis to movable ipsilateral axillary lymph node(s)
N2	Metastasis to ipsilateral lymph node(s) fixed to one another or to other structures
N3	Metastasis to ipsilateral internal mammary lymph node(s)
M	**Distant metastasis**
Mo	No distant metastasis
M1	Distant metastasis (including metastasis to ipsilateral supraclavicular lymph nodes)

Stage 0	Tis	No	Mo
Stage I	T1	No	Mo
Stage IIA	T0	N1	Mo
	T1	N1	Mo
	T2	N0	Mo
Stage IIB	T2	N1	Mo
	T3	N0	Mo
Stage IIIA	T0	N2	Mo
	T1	N2	Mo
	T2	N2	Mo
	T3	N1/N2	Mo

Stage IIIB	T4	Any N	Mo
	Any T	N3	Mo
Stage IV	Any T	Any N	M1

1.3.5. Treatment of the Breast Tumors

The goal of any oncologic treatment is to maximize the cure and at the same time optimize the quality of life [36].

1.3.5.1. Surgical Therapy

Surgical treatment represents most frequently used and the most successful sign method of cancer therapy currently available. More patients are cured of cancer by surgery than by any other therapeutic modality [37].

1.3.5.2. Conservation Breast Cancer Surgery

It is aimed at removing the tumor plus a rim of at least (1 cm) of normal breast tissue. This is commonly referred to as a wide local excision or lymphectomy [38].

1.3.5.3. Mastectomy

Removal of all breast tissue, choice may not be offered if the lesion is too large, multi-focal, and lobular or, in the surgeon's opinion, so close the nipple that it is likely to cause distortion [39].

1.3.5.4. Radiotherapy

Palliative radiotherapy may be advised for locally advanced cancers with distant metastasis in order to control ulceration, pain, and other manifestation in the breast and regional nodes [40]. Radiotherapy is especially useful in the treatment of the isolated bony metastasis, chest wall recurrence and brain metastasis [41,42].

1.3.5.5. Chemotherapy

Breast cancer is responsive to all major classes of cytoxic drugs: alkylating agents, antimetabolites, mitotic inhibitor, and the antitumor antibiotic[43]. Among the most active are alkylating agent including cyclophosphamide and thaitepa, and anthracyclines such as doxorubicin. The antimetabolites methotrexate (MTX) and 5–flourouracil (5-FU) are also active. Numerous combination of chemotherabutic agents have been evaluated in the treatment of metastatic breast cancers such as: CMF, CAF [44-47].

1.3.5.6. Endocrine Therapy

Hormonal therapy is the initial treatment of metastatic disease in patients with ER or PR positive tumors. Tamoxifen, a nonsteroidel antiestragon was approved first for the treatment of metastic breast cancer over 20 years ago, is usually the first agents of choice because of its favorable toxicity profile.[48,49]

1.3.6. Detection and Diagnosis (50)

Although an accurate history and clinical examination are still the most important method of detecting breast disease, there are a number of investigations that can assist in the diagnosis as follows:

1.3.6.1. Self Examination

All women over age 20 should be advised to examine their breast monthly–premenopausal women should perform the examination 7-8 days after the menstrual period. The breast should be inspected initially while standing before a mirror with the hands at the side, overhead, and pressed firmly on the hips to contract the pectoralis muscles. Masses asymmetry of breasts and slight dimpling of the skin may become apparent as a result of these maneuvers. Next, in a supine position, each breast should be carefully palpated with the fingers of the opposite hand. Some women discover small breast lumps more readily when their skin is moist while bathing or showering. Most women do not practice self-examination, and its value is controversial. Clearly, however, it is not harmful, it is inexpensive, and it may be beneficial [35].

1.3.6.2. Mammography

Soft tissue x-rays are taken by placing the breast in direct contact with ultrasensitive film and exposing it to low-voltage, high-amperage x-rays. The dose of radiation in approximately 0.1 Gy and therefore mammography is a very safe investigation [50].

1.3.6.3. Ultrasound

Ultrasound is particularly useful in young women with dense breasts in whom mammograms are difficult to interpret, and in distinguishing cysts for solid lesions. It can also be used to localize impalpable breast lumps [50].

1.3.6.4. Magnetic Resonance Imaging (MRI)

MRI is of increasing interest to breast surgeons in a number of settings, it can be useful to distinguish scar from recurrence in women who have had previous breast conservation therapy for cancer (although it is not accurate within 9 months of radiotherapy because of abnormal enhancement); it is the gold standard for imaging the breast of women with implants; it may prove useful as a screening tool in high-risk women; and it is being evaluated in the management of the axilla in both primary breast cancer and recurrent disease [50].

1.3.6.5. Needle Biopsy/Cytology

Histology can be obtained by using a fine needle such as a trucut or corecut biopsy device under local anesthesia. Cytology is obtained by using a 21 or 23 gauge needle and 10 mL syringe with multiple passes throughout the lump without releasing the negative pressure in the syringe. The aspirate is then smeared on to a slid, which is air-dried. Fine needle aspiration cytology (FNAC) is the least invasive technique to obtain a cell diagnosis and is very accurate if both operator and cytologist are experienced. However, false negatives do occur mainly through sampling error, and invasive cancer cannot be distinguished from in situ disease [50].

1.3.6.6. Triple Assessment

In any patients who presents with a breast lump or other symptoms suspicious of carcinoma, the diagnosis should be made by a combination of clinical assessment, radiological imaging and tissue sample taken for either cytological or histological analysis [50].

1.4.1. Definition of Tumor Marker (51)

A tumor marker is a substance present in or produced by a tumor or by the tumor's host in response to the tumor's presence that can be used to differentiate a tumor from normal tissue or to determine the presence of a tumor based on measurement in the blood or secretions. Such a substance can be found in cells, tissue or body fluids. It can be measured qualitatively or quantitatively by chemical, immunological, or molecular biological methods to identify the presence of a cancer.

Tumor markers are the biochemical for immunological counterparts of the differentiation state of the tumor. In general, tumor markers represents re-expression of substances produced normally in embryoginically closely related tissues. Few markers are specific for a single individual tumor (tumor-specific markers); most are found with different tumor of the same tissue type (tumor-associated markers). They are present in higher quantities in cancer tissue or in blood from cancer patients than in benign tumors or in the blood of normal subjects [51].

1.4.2. Routes of Tumor Markers Production

Benign tumors are generally well differentiated. The cells in a benign tumor are similar to the cells of the normal tissue, and the tumor markers produced are the products found in the normal tissue. They may be found in increased amounts in the circulation depending on the size of the tumor.

Malignant tumors may produce substance may associated with normal cell, or they may be different. As a zygote is transformed into an embryo, which then evolves into a fetus, the rapidly dividing cells became differentiated into specialized tissues by selective gene expression. The genes expressed are responsible for the production of hormones, enzymes, receptors, structural proteins, and cell metabolism. When a normal cell is transformed into tumor cell, gene expression changes. The affected cell may lose its ability to synthesize some specific cell products, or it may manufacture greatly increased amounts. The cell may be less specialized than the tissue it evolved from and assume the characteristics of the less well-differentiated cells of the embryo, synthesizing proteins found in the embryo but not in a normal adult. Cell proliferation rates change as the metabolic rate of the cells increases. After the cell is transformed, it loses growth control and begin to divide rapidly. The cells lose contact inhibition and invade the primary site. They then invade the adjacent organs and blood and lymph system, which may carry the cells to distance organs. The cell may then lodge in a capillary bed and begin to invade the new site. As this invasion process takes place, new proteins are produced that actively aid in the invasion. These proteins can also be used as markers. [52]

1.4.3. Classification of Tumor Markers

Tumor markers may be classified into chemical and genetic tumor markers. [52]

1.4.3.1. Chemical Tumor Markers

Table (1.3) summarizes the chemical tumor markers classified according to biochemical characteristics, and their associated malignancy. The table shows the low specificity of tumor marker for cancer [52].

Table (1.3): Chemical tumor markers [51].

Marker Type	Example	
Enzyme	Alcohol dehydrogenase	Liver
	Alkaline phosphatase	Bone, Liver, Leukemia, Sarcoma
	Alkaline phosphatase Placental	Ovarian, Lung, trophoplastic gastrointestinal, seminoma, Hodgkin's
	Amylase	Pancreas, Various
	Aryl Sulfatase B	Colon, breast
	Galactosyi transferase	Colon, bladder, gastrointestinal, Various
	Neuron-Specific enolase	Lung (small-cell), neuroblastoma, carcinoid, melanoma, Pheochromocytoma, pancreatic
	Prostate-specific antigen (PSA) Ribonuclease	Prostate Various (Large bowel. Lung, ovarian)
	Telomerase	Colorectal, Breast, etc.
	Sialyl transferenase	Colon, Breast, Lung
Hormone	ACTH	Gushing's syndrome, lung (small – cell)
	Antidiuretic hormone	Lung (small - cell) adrenal cortex, deudonal
	Calcitonin	Medullary thyroid
	Growth hormone hCG	Pituitary adenoma, renal, lung, Embryonal, choriocarcinoma, testicular (nonseminomatous)
	Human placental lactogen	Trophoblastic, gonads, lung, breast
	Parathyroid hormone	Liver, renal, breast, lung, various
	Prolactin	Pituitary adenoma, renal, lung. Breast
	Vasoactive intestinal	Pancreas, bronchogenic,
	Peptide 5	Pheochroinocytom neuroblastoma

Table (1.3): Continued.

	α-Feto protein	Hepato cellular, germ line (non-nseminoma)
	β-oncofeta antigen	Colon
	Carcino fetal ferritin	Liver
Oncofetal Antigen	CEA	Colorectal, gastrointestinal, pancreatic, lung, breast
	Pancreatic oncofetal	Pancreatic
	Sequamous cell antigen	Cervical, lung, skin, head and neck (squamous)
	Tennessee antigen	Colon, gastrointestinal, bladder
	CA 125	Ovarian, endometrial
	CA 15-3 (Episialin)	Breast, Ovarian
Mucin	CA 27-29	Breast
	MCA	Breast, ovarian
	Du-PAN-2	Pancreatic, ovarian, gastrointestinal, lung
	CA 19-9	Pancreatic, hepatic; gastrointestinal
Blood group related antigen	CA 19-5	GastrointestinaT, pancreatic, ovarian
	CA50	Pancreatic, colon, gastrointestinal
	CA 27.4	Ovarian, breast, colon, gastrointestinal
	CA 242 1	Ovarian, breast, colon, gastrointestinal
	β2-Microglobulin	Multiple myeloma, β-cell lymphoma Clironic lymphocytic
	C-peptide	Insulinoma
	Ferritin	Liver, lung, breast, leukemia
	Immunoglobuin	Multiple myeloma, lymphomas
Protein	Melanoma associated anfmen	Melanoma
	Pancreas associated antigen	Pancreatic, stomach
	Pregnancy specific protein	Trophopiastic, germ cell
	Prothrombin precursor	Meato cellular

	Tumor associated trypsin inhibitor	Ovarian
	Estrogen and progesteron receptors	Breast
	Catecholamine metabolites	Neuroblastoma, pheochromocytoma
Others	Hydroxy proline	Bone metastasis (breast) multiple myeloma
	Lipid-associated sialic acid	Gastrointestinal, lung. Rheumatoid
	Polyamine	Brain, various

1.4.3.2. Genetic Tumor Markers (51)

Two classes of genes are implicated in the development of cancer: Oncogenes (Cell activation genes–table 1.4) and suppressor genes (genes involved in the recognition and repair of damaged DNA-table 1.4). Oncogenes are derived from proto-oncogenes, which may be activated by dominate mutations. The type of mutation could be point mutation, insertion, deletion, translocation, and inversion. Most oncogenes code for proteins that function at the same stage of activation of cells for proliferation, and there activation leads to cell division. Most oncogenes are associated with hematological malignancies, such as Leukemia and to a lesser extent, solid tumors.

The other class of tumor genes the suppressor genes has been isolated from mostly solid tumors. The oncogenicity of suppressor genes is derived from the loss of the gene rather than their activation, as with oncogenes. The major tumor suppressor gene, P_{53}, functions to repair damaged DNA by apoptosis (programmed cell death).

Repair is mediated by activation of the production of P_{21}, which blocks the cell cycle in late G_1 to allow repair to take place. The loss of function of this gene may result in the inability of the DNA repair process and lead to the development of tumorgensis.[53]

The exciting promise of using detection of oncogenes and suppressor genes, for the diagnosis, determining the prognosis, and predicting the response to chemotherapy remains to be realized. However, oncogenes detection remains an experimental approach to human cancer, with great expectations not yet fulfilled. The ability to develop cancer by detection of mutations in tumor suppressor genes raises ethical questions that remain to be resolved.

Table (1.4): Classification of genetic tumor markers [52] .

Marker Type	Example	Associated Malignancy
Oncogene	N-ras mutation	Acute myeloid leukemia neuroblastoma
	K-ras mutation	Leukemia, lymphoma
	C-myc translocation	b-and T-cell lymphoma., small cell lung] carcinoma..

	C-erb B-2 amplification	Breast, ovarian, gastrointestinal
	C-abllber translocation	Chronic myelocytic leukemia
	N-myc amplification	Neuroendocrine
	bcL-2	Leukemia, iymphoma
Suppressor Gene	VHL mutation	Kidney
	APC mutation	Colorectal
	PI 6 (cd Kn2) mutation	Bladder, glioblastoma, melanoma
	WT1 mutation	Wilms', tumor
	Loss of heterozygosity	Wilms', breast, hepatoblastoma rhabdomyosarcoma, bladder
	BRCA2, PB1	Breast
	RB1 mutation	Retinoblastoma, osteosarcoma small-cell lung
	PI 6 E-cadheim mutation	Breast
	BRCA1 mutation	Neurofibromatosis 1 Melanoma, breast
	P_{53} mutation	Breast, colorectal, lung, liver, renal cell, bladder, sarcomas.
	DCC mutation	Colorectal
	NF_2 mutation	Neurofibro matosis2, meningioma

1.4.4. Clinical Applications of Tumor Markers (54)

The potential uses of tumor markers are summarized in table (1.5). In general, tumor markers may be used for diagnosis and prognosis of carcinomas and for monitoring the effects of therapy as well as targets for localization and therapy. Ideally, a tumor marker should be produced by tumor cells and be detectable in body which fluids should not be present in healthy people or in benign conditions. Therefore, it could be used for screening for the presence of cancer in symptomatic individuals in general population.

Most tumor markers are present in normal, benign, and cancer tissues. They are not specific enough to be used for screening cancer. In situations where the incidence of cancer is high among certain populations, screening might be possible.

Table (1.5): Clinical Usefulnees of tumor markers [55].

	Biochemical properties	Molecular weight	Primary clinical applications
Alpha-fetoprotein (AFP)	Glycoprotein, 4% carbohydrate; considerable homology with albumin	~70 KD	Diagnosis and monitoring of primary hepatocellular carcinoma and germ cell tumors. Prognosis of germ cell tumors.

Cancer antigen 125 (CA 125)	Mucin identified by monoclonal antibodies	~200 KD	Monitoring ovarian carcinoma. Prognosis after chemotherapy
Cancer antigen 15-3 (CA 15.3, BR 27.29)	Mucin identified by monoclonal antibodies	>250 KD	Monitoring breast cancer
Cancer antigen 72.4 (CA 72.4)	Glycoprotein identified by monoclonal antibodies	~48 KD	Monitoring gastric carcinoma
Cancer antigen 19-9 (CA 19-9)	Glycolipid carring the Lewisa blood group determinate	~1,000 KD	Monitoring pancreatic carcinoma
Carcinoembri-yonic antigen (CEA)	Family of glycoproteins, 45%-60% carbohydrate	~180 KD	Monitoring gastrointestinal and other adenocarcinomas
CYFRA 21-1	Fragments of cytokeratin	~30 KD	Monitoring bladder and lung carcinoma
Estrogen receptor	Nuclear transcription	65 KD	Predicting response to endocrine therapy in breast cancer

Table (1.5): Continued.

Human chorionic gonadotrophin (hCG)	Glycoprotein hormone consisting of tow non-covalently bound subunits (α and β)	~36 KD	Diagnosis and monitoring non-seminomatous germ cell tumors, choriocarcinomas, hydtidiform moles, seminomas. Prognosis of germ cell tumors.
Neuron specific enolase (NSE)	Dimer of the enzyme enolase	~87 KD	Monitoring small cell lung carcinoma, neuroblastoma, apudoma.
Placental alkaline phosphatase (PLAP)	Heat-stable isoenzyme of alkaline phosphatase	~86 KD	Monitoring of germ cell tumors (seminomas)
Progesterone receptor	Nuclear transcription factor	A from: 94 KD B from: 120 KD	Predicting response to endocrine therapy in breast cancer .
Prostate specific antigen (PSA)	Glycoprotein serine protease	~36 KD	Diagnosis, screening and monitoring prostatic carcinoma

Squamous cell carcinoma antigen (SCC)	Glycoprotein sub-fradion of tumor antigen T4	48 KD	Monitoring squamous cell carcinomas
Tissue polypeptide antigen (TPA)	Fragments of cytokeratin 8,18 and 19	~22 KD	Monitoring bladder and lung carcinoma
Tissue polypeptide specific antigen (TPS)	Fragment of cytokeratins 18	~22 KD	Monitoring metastatic breast carcinoma

1.4.5. Tumor Markers in Breast Cancer (56)

Several tumor markers have been investigated for one or more clinical use in breast cancer (Table 1.6).

Tumor-associated antigens (TAAs) that have been associated with breast cancer include carcinoembryonic antigen (CEA); tissue polypeptide antigen (TPA), tissue polypeptide-specific antigen (TPS), gross cystic disease protein (GCDP); prostate specific antigen (PSA); and the products of the MUC-1 gene. The MUC-1 gene encodes a cell-associated mucin-like protein. Secretary epithelial cells such as breast epithelial cells express this antigen. Several assays detect the MUC-1 gene products, but they are not identical. These proteins have been identified by monoclonal antibodies to breast cancer cell lines, breast cancer tissue, or human milk fat globule membranes. Assays that detect circulating MUC-1 products include CA 15-3, CA 27-29, CA

549, breast cancer mucin (BCM), mammary serum antigen (MSA), and mucin-like carcinoma-associated antigen (MCA) [57].

The results obtained with these assays may not be identical, presumably due to reactivity of different antibodies to different epitopes, and/or different sensitivities and specificities that result from different assay configurations [46].

More recently, markers of tumor biology have been investigated in breast cancer (Table 1.6), and molecules related to angiogenesis, adhesion, invasion, and metastases. Several, but not all of these are indeed, detected with immunologic assays, and could arguably be designated as TAAs.

Table (1.6): Tumor markers that have been investigated in breast cancer[56]

Tumor-associated antigens
Carcinoembryonic antigen (CEA)
Products of or related to products of the MUC-1 gene
CA 15-3
CA 27-29
CA 549
Breast cancer mucin (BCM)
Mammary serum antigen (MSA)

Table (1.6): Continued.

Mucinous carcinoma antigen (MCA)
Tissue polypeptide antigen (TPA)
Tissue polypeptide-specific antigen (TPS)
Gross cystic disease protein (GCDP)
Prostate-specific antigen (PSA)
Markers of tumor biology
Extra-ceullular domain (ECD) of c-erbB-2/HER2/neu
Molecules of adhesion and invasion
E-selectin
Soluble urokinase plasminogen activator receptor (SuPAR)
Intercellular adhesion molecule-1 (ICAM-1)
Molecules associated with angiogenesis
Vascular endothelial growth factor (VEGF)
Basic fibroblast growth factor (bFGF)
Hepatocyte growth factor (HGF)
HUVEC assay
Antibody response against TAAs
c-erbB-2/HER2/neu
P_{53}

There are several tumor markers correlate with the incidence of breast cancer, but the most important markers are:

1.4.5.1. CEA

Carcinoembryonic antigen is a marker for breast carcinoma [59], lung, gastrointestinal and colorectal [60]. CEA is one of the older oncofetal protenis in use. CEA is a large family of related cell-surface glycoproteins with a high molecular mass of 150 to 300 KD, it contains 45 to 55% carbohydrate with increase expression found in a variety of malignancies, including breast cancer[61] .CEA is not recommended for screening, diagnosis, staging or routine surveillance of breast cancer patients following primary therapy [62].

1.4.5.2. TPA

Tissue polypeptide antigen is not a specific tumor marker [63]. Antibodies that react with cytokeratin 8,18 and 19 identify it. TPA is a heterogeneous group of molecules with molecular weight range 20-45 KD [64]. Both normal and cancerous cells produce TPA; it is useful in the monitoring of metastic diseases[54].

1.4.5.3. TPS

Tissue polypeptide-specific antigen (TPS) is a new tumor marker defined by monoclonal antibody against the soluble tissue polypeptide antigen (TPA) [65]. First described as specific tumor marker by Bjorklund in 1957 [63]. In breast cancer patients TPS was especially useful in monitoring response to treatment and effectiveness of therapy in metastatic disease [66].

1.4.5.4. CA 549

CA 549 is an acidic glycoprotein and it is a marker for breast carcinoma. CA 549 is not useful in detecting early breast carcinoma but it is useful is detecting recurrence of breast cancer in patients after initial therapy followed by adjuvant therapy [67].

1.4.5.5. CA 27.29

CA 27.29 is detected by a monoclonal antibody B 27.29 [68], this is produced against antigen in ascites of patients with metastatic breast carcinoma. CA 27.29 test above 37.7 KU.L^{-1} were considered positive, its most useful in monitoring metastatic breast carcinoma [69].

1.4.5.6. CA 125

CA 125 is a high-molecular mass (>200 KD) glycoprotein recognized by the monoclonal antibody OC 125. The level of CA 125 is measured quantitatively by using immunoradiometric assay [70]. In healthy population, the upper limit of CA 125 level is 35 KU.L^{-1}. CA 125 is elevated in ovarian carcinoma, endometerial, pancreatic, Lung, breast, colorectal and other gastrointestinal tumors [71,72]. CA 125 is useful to detecting residual disease in cancer patients following initial therapy [73].

1.4.5.7. Mammary Antigen

Several new antigens have been recognized by monoclonal antibodies. Which have been identified in patients with breast cancer [74]. They have been proposed as "tumor markers":

• MCA

Mucin-like carcinoma associated antigen (MCA) is a mucin glycoprotein with a molecular mass of 350 KD. MCA was identified on the surface of a breast carcinoma cell line by the monoclonal antibody b-12 [54]. MCA level is elevated in 60% of metastatic breast cancer patients [75].

• MAM-6

MAM-6 an epithelial membrane antigen present on ductal and alveoli epithelial cells that is detected by monoclonal antibody raised against human milk-fat globule membranes [76]. Partial characterization of the antigen by SDS-PAGE showed that the antigen is a polymorphic epithelial sialomucin with a molecular mass over 400 KD [77].

• MSA

Mammary serum antigen (MSA) was detected by an antibody raised against a whole cell suspension of intraductal breast cancer [74].

1.4.5.8. Galectin-4

A protein Galectin-4 is expressed in non-invasive and invasive breast cancer but not in normal cell. An anti-Galectin-4 antibody was able to detect the presence of Galectin-4 very specifically. Galectin-4 is specific diagnostic marker of breast cancer whose patterns of expression at early stages of disease could identify those patients with a high risk of progression to aggressive cancer [78].

1.4.5.9. Cathepsin-D

Cathepsin-D is a glycoprotein with molecular weight M.wt: 52 KD. It was discovered in 1979 in the culture medium of hormone dependent human breast cancer. It is a precursor to lysosomal acidic protease. This proteolytic enzyme can react against basement membranes [79].

Cathapsin-D may facilitate cellular actions such as migration, metastasis, and an invasion of other tissues. Estrogen has been shown to stimulate secretion of this tumor marker in certain hormone-dependent breast cancer cell lines. This antigen has been found to have potential application in breast cancer prognosis as its concentration appears to be related to the patients overall change for survival [80,81].

CA 15-3 is a breast-associated antigen identified on the apical side of alveoli and ducts of mammary glands and as a circulating antigen [82]. Distinct epitopes of this high molecular-weight mucin-like glycoprotein of 300-400 KD[83-85], which carbohydrate side chain account for about 50% [86].

Also known as polymorphic epithelial mucin [87] (PEM), epithelial membrane antigen [88] (EMA) or episialin [89]. CA 15-3 can be identified by two monoclonal antibodies DF3 and 115 D8, in a double-determinate or sandwich-type immunoassay [90]. The 115 D8 antibody was prepared against human milk-fat globulin membrane [91] while the DF3 antibody was raised against a membrane-enriched fraction of a human breast carcinoma [92].

1.5.1. Structure of CA 15-3

CA 15-3 (Episialin) is synthesized as transmembrane molecule with a relatively large extracellular domain and cytoplasmic domain of 69 amino acids[93]. The extracellular domain mainly consists of region of nearly identical repeats population, leading to substantial differences in molecular weights of the CA 15-3 molecules from different individuals [94].

The repeats together with adjacent degenerated repeats contain many serins and threonines that are potential attachment sites for O-liked glycans and constitute the mucin-like domain, which comprises more than half of the polypeptide backbone. The mucin domain of CA15-3 contains many prolines and other helix-breaking amino acids, resulting in a molecule with an extended structure and many β-turns [95]. The extended structure is very rigid as aresult of the numerous O-linked glycans attached to the molecule [96]. The CA15-3 extends 200 to 500 nm above the cell membrane [97].

1.5.2. CA 15-3 Expression

- **CA 15-3 Expression in Normal Tissues**

CA 15-3 is predominatly found at the apical side of epithelial cells lining the acini alveoli, or lumens in various organs, i.e. in the mammary glands, salivary glands, sebacious glands, sweat glands, esophagus, stomach, pancreas, bile ducts, lungs, kidney, bladder, prostate, uterus, and rete testis [98-100].

- **CA 15-3 Expression In Malignant Tissues**

Relative to the expression levels of CA 15-3 found in normal tissues, CA15-3 is often overexpressed several-fold in many types of carcinomas derived from these tissues [101]. In these tumors, polarization of the cells is often lost, resulting in the presence of CA 15-3 at the entire cell surface. High levels of CA 15-3 are also detected on carcinoma cells present in pleural effusions on ascites from patients with breast or ovary carcinoma and on many breast carcinoma cell lines.

1.5.3. Biosynthesis of CA 15-3

CA15-3 is synthesized as a large single polypeptide, in most cell lines approximately 200 KD or more [102,103]. This precursor is rapidly cleaved by proteolysis in a small moiety, which contains the transmembrane and cytoplasmic domains, and a larger part, which comprises most of the extra cellular domain. Both moieties remain non-covalently associated [104]. This proteolytic processing step occurs in the endoplasmic reticulum and may be essential for further maturation. CA 15-3 is mainly processed by adding numerous O-linked glycans, which increases the apparent molecular weight on SDS-polyacrylamide gels to more than 400 KD. The extensive glycosylation protects the molecule against proteolytic degradation, since the precursors without O-linked sugars are degarded rapidly, while the mature molecule is extremely resistant to the action of proteases. The glycosylation also determines the rigidity of the molecule. The last step in the processing of CA15-3 is the addition of sialic acid to the glycans, which increases the mobility of the molecule on SDS-gels [96].

The early proteolytic cleavage step is not directly responsible for the release of CA15-3 for the membrane, which suggests that CA15-3 is most likely released from the membrane by a second proteolytic cleavage step after arrival at the cell surface. The second proteolytic cleavage seems to be a slow and probably a random process, allowing the mucin to remain associated with the cell surface with a half-life of 16-24 hrs [96].

1.5.4. Methodology

The CA 15-3 test from all sources uses both DF3 and 115-D8 antibodies. Serum is initially incubated with a polystyrene bead to which 115-D8 antibody has been attached. This antibody binds to antigenic sites on the glycoprotein, pulling it out of solution. The beads are then washed to remove unbound meterial and incubated with the radioiodine (^{125}I)-labeled DF3 antibody. The radiolabeled DF3 antibody binds its antigenic sites and then the amount of radioactivity is quantitated [105]. This is called Immunoradiometric Assay (IRMA).

1.5.5. Biology of CA15-3

• **CA15-3 and Cell Adhesion**

Similar to mucins in mucus, membran-associated mucins might act as barrier molecules to protect cells against toxic substances, as in pancreatic and bladder ducts. The high densities of CA15-3, due to its extended and relatively rigid structure, might also interfere with the function of the adhesion molecules. In this way, CA15-3 might prevent interactions between opposing apical membrane of polarized normal cells and facilitate the formation and maintenance of the ducts during development [106,107].

In carcinomas, the combination of overproduction and loss of polarization of CA15-3 expression might reduce cell adhesion and facilitate the invasion of tumor cells because CA15-3 might now interfere with the function of molecules required from tissue integrity [104].

- **CA15-3 and Immune System**

    The putative function of CA15-3 in tumor progression may not only be restricted to inhibition of adhesion which will probably result in an increased invasive potential of cells, but CA15-3 overexpression may well be critical to the survival of tumor cells during dissemination [108]. A completely different aspect of CA15-3 is its ability to act as a tumor-specific antigen. The underglycosylation of CA15-3 in various tumor cells exposes the protein backbone, leading to the generation of novel epitopes. This could elicit an immune response [109-112].

1.5.6. Clinical Application

    In healthy subjects, the upper limit of CA 15-3 concentration is 25 ($KU.L^{-1}$). At this level, (5.5%) of 1050 normal subjects, (23%) of patients with primary breast cancer, and (69%) of those with metastatic breast cancer show elevated CA 15-3 levels [113].

    Elevated CA15-3 levels are also found in other malignancies, including pancreatic (80%), lung (71%), breast (69%), ovarian (64%), colorectal (63%) and liver (28%) cancer. It is also reported to be elevated in benign diseases, although with less frequency (e.g., in benign liver [42%] and benign breast diseases [16%]).

CA15-3 should be used to diagnose primary breast cancer, because the incidence of elevation (23%) is fairly low. CA 15-3 is most useful in monitoring therapy and disease progression in metastatic breast cancer patients. A significant change must be at least (25%) and correlates with disease progression in (90%) of patients, with its regression in (78%). No change correlates with disease stability in (60%). CA 15-3 could replace CEA in metastatic breast cancer owing to its sensitivity and specificity.

1.6. Carbohydrate Antigen 19-9 (CA 19-9)

1.6.1. Marker Definition

CA 19-9 is a carbohydrate antigen identified as a glycolipid-that is, sialylated lacto-N-fucopentose II ganglioside, which is a sialylated derivative of the Lewis a blood group antigen and is denoted as Le ^{a} [114]. CA19-9 is synthesized by normal human pancreatic and biliary ductular cells and by gastric, colonic, endometerial, kidny, salivary gland, sweat gland and present in ductal epithelium of breast [115-117]. In serum it exists as a mucin, a high-molecular weight (200-1000 KD) glycoprotein complete [54]. The monoclonal antibody against CA19-9 was developed from a human colon carcinoma cell line, SW-1116 by Koprowski and associates [118].

Monoclonal antibody 19-9 derived from spleen cells of a mouse immunized with human colon adenocarcinoma cell line SW-1116 [119]. The epitope of this antibody is carbohydrate with the sugar sequence

NeuNAcα 2-3 Gal β 1-3 GlcNAc β 1-3 Gal...

$$4$$

$$|$$

Fuc α 1

As described by Magnani et.al. [119].

1.6.2. Methodology

CA 19-9 is measured with a double monoclonal immunoradiometric assay, using monoclonal antibodies raised against the SW-1116 cell line [120]. The antibody reacts with CA19-9 found at low concentrations in sera from healthy individuals, but frequently increased in sera from patients with adenocarcinomas [120]. The upper limit of normal for healthy subjects has been defined by the cutoff value of 37.0 (U.mL^{-1}) [121]. CA 19-9 has become an established marker for pancreatic cancer [121-123], but it must still be regarded as a research test for colorectal cancer.

Another methods to determinate CA 19-9 were enzyme-linked immunosorbent assay. Both the capture and the enzyme-conjugated antibody use the CA 19-9 monoclonal antibody. It should be noted that this antibody is useless for cancer diagnosis when a patient is lacking the enzyme for the synthesis of sialyl Le [a]. In Japanese, about 5-10% of the population lacks this enzyme. Determination carbohydrate antigen CA 19-9 levels in serum were also measured by radioimmunoassay (RIA) [125]. Immunohistochemical technique used for the distribution of CA19-9 in tissues using an immunoperoxidase assay [126].By this technique the CA 19-9 can be detected not only in cancerous tissues but also in non cancerous normal tissues.

1.6.3. Screening

Numerous studies have addressed the potential utility of CA 19-9 in adenocarcinoma of the colon and rectum.

The reported incidence of elevated serum CA 19-9 in colorectal cancer ranges from 20% to 40% [127,128]. The incidence of elevated CA 19-9 in stage-related, with the highest sensitivity occurring in patients with metastases [129-

[131]. However, the sensitivity of CA 19-9 was always less than that of the CEA test for all stages of disease [127-130]. The false-positive rate (>37.0 $U.mL^{-1}$) is 15% to 30% in patients with non-neoplastic diseases of the pancreas, liver and biliary tract [131]. Consequently, CA 19-9 cannot be used for screening asymptomatic populations.

1.6.4. Monitoring Response to Treatment

Kouri et.al.[132] compared CEA and CA 19-9 for predicting response to chemotherapy in 85 patients. Decreases in CEA more accurately reflect the response to therapy than did the decreases of CA 19-9. The pretreatment CA 19-9 value was, however, an important prognostic factor. Median survival was 30 months for patients with normal CA 19-9 values and 10.3 months for patients with elevated CA 19-9 values. CA 19-9 used to examined the serum levels and immunohistochemistry during the clinical course of female patient treatment with idiopathic interstitial pneumonia (IIp) that had elevated serum levels of   CA 19-9 [133].

1.6.5. Clinical Application

Elevated levels (>37 U.mL^{-1}) were seen in patients with pancreatic (80%), hepatobiliary (67%), gastric (40-50%), hepatocellular (30-50%), colorectal (30%), and breast (15%) cancer. Pancreatits and other benign gastrointestinal diseases show a 10 to 20% elevation; however, the levels are usually lower than 120 (U.mL^{-1}). CA 19-9 levels correlate with pancreatic cancer staging [54]. CA19-9 is useful in monitoring pancreatic and colorectal cancer. Elevated levels can indicate the recurrence before clinical finding by 1 to 7 months [134]. Unfortunately, early detection of relapse may not be useful because of the lack of effective therapy for pancreatic cancer.

ntroduction

The role of tumor markers in breast cancer is to enhance the clinicians, ability to provide more effective management of the disease [135]. Serum CA15-3 concentration was determined by using sandwich enzyme immunoassay of a double monoclonal antibody [136,137], automated chemiluminescent immunoanalyzer [138], immunoradiometric assay [139] and radioimmunoassay[140], in women with benign breast tumor and breast cancer.

CA 15-3 has been used in management of patients, with breast cancer. CA 15-3 has been evaluated for its ability to determine diagnosis, prognosis, monitor therapy and predict recurrence of breast cancer following curative surgery and radiation therapy [141,142]. Low incidences of CA 15-3 elevation in early stage cancer (stage I and stage II) have been observed [143].

Incidence of abnormal values of CA 15-3 in stage III and stage IV, and a very high CA 15-3 level have been correlated with metastases of breast cancer[144].

Therefore the development of immunoradiometric was planned to carry out the determination of the optimum conditions of ^{125}I-anti CA 15-3 antibody binding with CA 15-3 in breast tumor tissue homogenate, hence determination of CA 15-3.

Chapter Two

CA15-3 in Breast Tumors

2.1. Materials

2.1.1. Chemicals

All chemicals and reagents used in this study were of analytical grade, tabulated in the following table.

Table (2.1): Chemicals used and Companies.

Chemicals	Company
1. Immunoradiometric assay kit for CA 15-3 level	Diasorin Inc. (USA)
2. Bovine serum albumin (BSA) , urea , $ZnCl_2$,$CaCl_2$,NH_4Cl, NaBr, ethylendiamine-tetraaceticdisodium salt (EDTA).	Fluka: (Switzerland)
3. $CuSO_4.H_2O$, NaK-tartarate glycine, NaOH,HCl, $NaCO_3$,NaF,NaCl,NaI,Na_2HPO_4, NaH_2PO_4.	BDH,limited,Poole (UK)
4.Folin-Ciolteau	E.Merck AG. Dastmstapt
5.Blue dextran (2000),sepharose CL-4B.	Pharmacia fine chemicals (Sweden)

2.1.2. Instruments

Table (2-2): Instruments used and Companies.

Instruments	Company
1.Gamma counter type 1270-rack gamma II 2. Spectrophotometer ultraspace type 4050	LKB
3. UV-210 a double beam spectrophotometer	Shimadzu

4.pH-meter	**Pye-Unicam**
5.Cooling centrifuge; with a maximum speed 5000 r.p.m.	**Hettich**
Cooling centrifuge type 202-MK; with a maximum speed 13500 r.p.m.	**Sigma**
7.Memmert water bath, memmert incubator	**West Germany**
8. SM-shaker	**England**
9. Combicold rack	**LKB**

2.1.3. Patients

Three groups of breast tumors patients were included in this study.

Group I : Consisted of 40 patients with benign breast tumors

Group II : Consisted of 32 premenopausal patients with breast cancer.

Group III : Consisted of 15 postmenopausal patients with breast cancer.

Group IV : Consisted of 10 controls.

All patients were admitted for treatment to (Saddam Medical City, Baghdad Teaching Hospital),(University Hospital, Saddam College of Medicine), (Nursing Home Private Hospital) and (Al-Arabi Private Hospital).

Patients suffered from any disease that may interfere with this study were excluded. All surgical operation of breast tumors were carried out under the supervision of the following surgeons:

Dr. Saab Sedeq, Dr.Munthir Al-Aubaidi, Dr.Azam Qanbar Agha, Dr. Abd Al-Salam Al-Tai, Dr. Zuhair Abid Al-Hadi.

The host information of all patients and normal healthy subjects is summarized in table (2-3).

Table (2.3): The host information of breast tumors patients and healthy subjects studied.

Group	Patients	No.	Age	Type of tumor	Metastases
I	Benign breast tumor	40	18-42	23 fibroepithelial tumor (fibroadenoma) 17 fibrocystic changes (adenosis)	– –
II	Premenopausal malignant breast tumor	32	34-52	22 Infiltrative Ductal carcinoma 10 Ductal carcinoma	2 lymph nodes
III	Postmenopausal malignant breast tumor	15	55-73	Infiltrative Ductal carcinoma	4 lumph nodes
IV	Control	10	25-40		

2.1.4. Preparation of Blood Samples

Five milliliters of blood samples were obtained from patients by venipuncture just before surgery. Ten physically normal age volunteers were used as controls. Blood samples were left for 20 min. at room temperature. After coagulation, sera were separated centrifugation at 3000r.p.m for 10 min., and then sera were aspirated and stored at −20°C until time analysis. The samples were not thawed and refrozen before testing.

2.1.5. Collections of Specimens

The tumors tissues were surgically removed from breast tumor patients by either mastectomy (cancer patients) or lumpectomy (benign tumor patients). The specimens were cut off and immediately rinsed with ice-cold isotonic saline solution. They were collected individually in plastic receptacle and stored at $-20\,^{\circ}C$ until homogenization.

2.1.6. Preparation of Phosphate–Buffered Saline

Phosphate –buffered saline (PBS) 0.15 M, pH 7.2 was prepared as following:

A: Disodium basic phosphate (0.15M); 21.2940g Na_2HPO_4 and 9.0g of NaCl were dissolved in a final volume 1L deionized distilled water.

B:.Monobasic sodium phosphate (0.15M) 17.9970g of NaH_2PO_4 and 9.0g NaCl were dissolved in a final volume 1L deionized distilled water.

Phosphate buffer saline pH 7.2 was prepared by mixing a volume of solution A with appropriate amounts of solution B to obtain the required pH.

2.1.7. Preparation of Breast Tumors Tissues Homogenates

The frozen tissue were weighed, sliced finely and scalped in petri dish standing on ice bath, and then homogenized with fivefold volumes of PBS buffer pH7.2, using manual homogenizer [145]. The homogenate was filtered through four layers of nylon gauze in order to eliminate fibers connective tissues, and then centrifuged at 4000 r.p.m for 45 min. at 4 °C in order to precipitate the remaining intact cells and the intact nucleus. The supernatant fraction at this speed was separated, divided in aliquots and freezed-20 °C until use.

2.1.8. Statistical Analyses

Students' t-test was used to determine if the mean values of studied parameters were significant different in the individual groups included in this work. $P<0.05$ were considered significant [146].

2.2.1. Protein Determinations

Total homogenate protein content was determined by the method of Lowry [147], using bovine serum albumin (BSA) as the standard.

Figure (1.1) represents the standard curve of protein, which was constructed by plotting the absorbance at 600 nm against standard protein concentrations

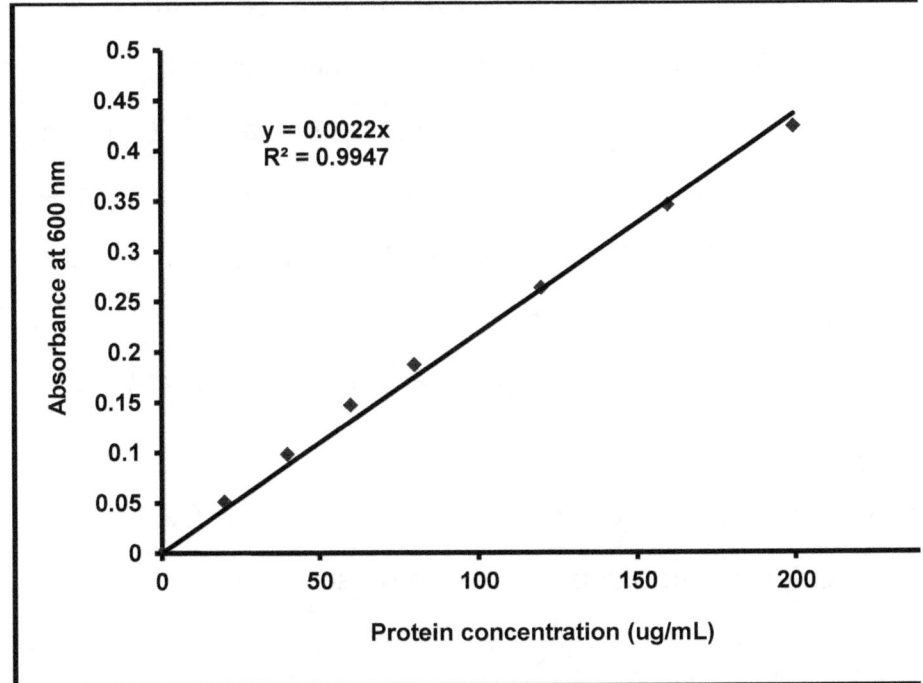

Figure (2.1): Standard curve of protein concentration. (All other details are explained in the text).

2.2.2. Determination of CA 15-3 Levels in Sera of Breast Tumors Patients

Reagents

The following reagents provided in the CA15-3 IRMA kit from Dia–Sorin-U.S.A. were used:

- Tracer: two vials each one contained 1.0 μ Ci/mL (37.1 KBa /mL). CA15-3 antibody labeled with ^{125}I in 10 mL / Tris buffer with protein stabilizer and preservative.

- CA15-3 standards: The vial contained 100 mL, which represented 0 U.mL^{-1}. There are four vials, 1.2 mL in each vial with different concentrations of human CA15-3 (25, 50, 100, 200) U.mL^{-1} in Tris. buffer with protein stabilizer and preservative.

- One bottle contained 100-coated beads, Anti-CA15-3-mouse, monoclonal.

- One vial contained 0.5mL CA15-3 control, CA15-3 in re-calcified human plasma with preservative.

Procedure

The assay protocol is described in table (2.4).

Table (2.4): IRMA protocol of serum CA 15-3 (U.mL^{-1}) (All other details are explained in the text).

	CA 15-3 standard (U.mL^{-1})					Control	Unknown samples	
	0	25	50	100	200		1	2-etc.
Reaction trays no.	1.2	3.4	5.6	7.8	9.10	11.12	13.14	15.16- etc.
Standard (µL)	200	200	200	200	200	–	–	–
Control serum or samples (µL)	–	–	–	–	–	200	200	200
^{125}I-anti CA15-3	200	200	200	200	200	200	200	200

The specimens and reagents must be brought to room temperature

(20-30 °C) before opening. The reaction trays and data sheets were marked.

First Incubation

The specimens and the control were diluted to (1:15) prepared by adding 20 µL of specimen or control to 1000 µL CA15-3 standard, 0 U.mL^{-1} in a tube marked proper identification of specimen. Two hundred microliters of diluted specimen and control were pipette to their assigned wells. Two hundred microliters of each standard was pipette to its assigned well (standards are not to be diluted). One bead was dispensed into each well and the adhesive cover sealer was applied. After incubation for 2hrs at room temperature, the adhesive cover sealer was removed and the liquid was aspirated, then each bead was washed three times with 5 mL distilled water.

Second Incubation

Two hundred microliters of ^{125}I-antiCA15-3 was pipetted on each bead. The adhesive cover sealer was applied again. After incubation time for 3hrs at room temperature, the cover was removed and the liquid was aspirated from wells, then the beads were washed as it is above. The beads were transferred to the counting tubes, and then the tubes were counted for 1 min.

Calculations

The standard curve was constructed by plotting counts per min. (Y axis) versus concentration for CA15-3 standard (X axis), figure (2-2). Then the points were connected with straight-line segments.

The CA15-3 concentration of specimens and control were determined directly from the standard curve.

2.2.3. Preliminary Test of CA 15-3 Binding to 125I -Anti CA15-3 Antibody in Breast Tumor Homogenate

Reagents

Phosphate buffered saline 0.15 M; pH 7.2 was prepared as described in section (2.1.6).

Procedure

The supernatant and pellet were centrifuged and detected by using ordinary tubes. In order to detect CA 15-3, 100 µL of the supernatant breast homogenate having (900µg protein) were incubated with 50 µL (0.35 mg.mL^{-1}) of ^{125}I - anti CA15-3.The volume of reaction was completed to 500 µL with PBS buffer pH7.2, then incubated at 37 °C for 2hrs. The assay tubes were centrifuged at 4000 r.p.m. for 45 min. at 4 °C.

The supernatant was discarded, the rim at each tube was swabbed with cotton, and then gamma counter counted the complex formed for one minute. The pellet of CA 15-3 was estimated by dissolving the sediment in PBS-buffer pH 7.2 with the ratio 1:5 (weight: volume) shaking was then carried out. Hundred microliters of the supernatant fraction of the sediment having (540 µg.mL^{-1} protein) was added to 50 µL (0.35 mg.mL^{-1}) of ^{125}I -anti CA 15-3 antibody. The same steps mentioned in this experiment were followed to determine the radioactivity of the complex formed. For total count two additional tubes with 50µL of ^{125}I –anti CA 15-3 antibody were counted in gamma counter.

Calculations

1. The counted radioactivity in each tube (expressed in c.p.m.) represents the bound fraction (B), (i.e., ^{125}I antiCA15-3 antibody/CA 15-3 complex).
2. The counted radioactivity in the tubes containing ^{125}I-anti CA15-3 antibody only represents the total count (T).
3. The (B/T) ratio for each tube counted as follows:

$$(B/T)\% = \frac{Sample\ Counts\ (B)}{Total\ Counts\ (T)} \times 100$$

2.2.4. Factors Effecting of 125I-Anti CA-3 Antibody Binding to CA 15-3 in Breast Tumors Homogenates

2.2.4.1. The Effect of Different Amounts of Protein Concentration of the Tumor Homogenate on the Binding with 125I-Anti CA 15-3 Antibody

Reagents

All reagents prepared as described previously in sections (2.1.6) and (2.2.3).

Procedure

1. Fifty microliters (0.35 mg.mL^{-1}) of ^{125}I -anti CA 15-3 antibody were added to 100µL of the supernatant (benign Fibroadenoma, pre-and post-menopausal malignant breast tumors (IDC) respectively) containing increasing amounts of protein (50, 100, 150, 200, 250 µg.mL^{-1}) then completed to a final volume of reaction to 500 µL with 0.15 M PBS pH 7.2.

2. The assay tubes were then incubated for 2 hrs at 37°C.

3. Two additional tubes, containing 50μL (0.35 mg.mL^{-1}) of ^{125}I –anti CAB-3 antibody only, for total counts were set-aside until counting.

4. At the end of incubation, the assay tubes were centrifuged at 4000 r.p.m for 45 min at 4°C.

5. The supernatant were decanted, the rims at the tube were swabbed with cotton piece.

6. The radioactivity of the complex were counted using gamma counter.

Calculations

1. The B/T percent were determined according to section (2.2.3).

2. The percent of binding values B/T were plotted versus the increasing amount of protein of the breast tissue homogenate.

2.2.4.2. The Effect of 125I -Anti CA15-3 Antibody Concentration on the Binding

Reagents

All reagents prepared as described previously in section (2.1.6) and (2.2.3).

Procedure

1. Fifty microliters of increasing concentration (0.070,0.140,0.175,0.350, 0.701 mg.mL^{-1}) of ^{125}I -anti CA 15-3 antibody were added to 100μL of homogenate (benign breast tumor (Fibroadenoma), pre-and post-menopausal malignant breast tumors) (IDC) containing (100, 100, 200 μg.mL^{-1} protein) respectively.

2. The volume of reaction was made up to 500 μL with PBS pH 7.2.

3. Steps 2,3,4,5 and 6 of the experiment (2.2.4.1) were repeated.

Calculations

1. The same mathematical equation mentioned in section (2.2.4.1) was used to calculate (B/T)%.

2. Values of (B/T)% were plotted versus concentration of labeled antibody (^{125}I -anti CA15-3 antibody).

2.2.4.3. The Effect of pH on the Binding

Reagents

All reagents prepared as described previously in section (2.1.6) and (2.2.3).

Procedure

1. One hundred microliters of human homogenate (benign breast tumor (Fibroadenoma), pre-and post-menopausal malignant breast tumors (IDC)) containing (100,100,200, $\mu g.mL^{-1}$ protein) were added to 50μL (0.175, 0.175,0.140mg.mL^{-1}) of ^{125}I -anti CA15-3 antibody respectively.

2. Each mixture was completed to 500 μL with PBS of different pH ranging (6.8-8.0).

3. Step 2,3,4,5 and 6 of the experiment (2.2.4.1) were repeated.

Calculations

1.Values of (B/T) % were calculated as described in section (2.2.4.1).

2. (B/T)% were plotted against their corresponding pH.

2.2.4.4. Time Course of the Binding of 125I -Anti CA15-3 Antibody to CA 15-3 in Breast Tumors Homogenates

Reagents

All reagents prepared as described previously in sections (2.1.6) and (2.2.3).

Procedure

1. One hundred microliters of homogenate (benign breast tumor (Fibroadenoma), pre-and post-menopausal malignant breast tumors (IDC)) containing (100,100,and 200 $\mu g.mL^{-1}$ protein) were incubated with 50μL of ^{125}I - anti CA15-3 antibody concentration (0.175,0.175 and 0.140 gm.mL^{-1}).

2. The volume of reaction was made up to 500 μL with PBS pH (7.0,7.6 and 7.8).

3. All tubes were incubated at 37°C at different time intervals (30, 60, 90, 120, 150 and 180) min.

4. Step 3,4,5,6 of the experiment (2.2.4.1) was repeated.

5. To determine the time course of CA 15-3 binding to ^{125}I – anti CA 15-3 antibody at different temperatures. Steps 1, 2, 3 and 4 in the same experiment were repeated at different temperatures (5, 15, 25, 45°C).

Calculations

1. The values of (B/T)% were calculated as described in section (2.2.4.1) at each time and temperature used.
2. The values (B/T)% was plotted against the time of incubation at different temperatures.

2.2.4.5. The Effect of Different Halides on the Binding

Reagents

1. Phosphate buffer (PB) were prepared as described in section (2.1.6) without the addition of NaCl .
2. Halid reagents were prepared in concentration of 0.01M PB at pH (7.0, 7.6 and 7.8) individually, by dissolving each of 0.021gm of NaF, 0.0292gm of NaCl, 0.0515 gm of NaBr, and 0.075gm of NaI in a final volume 50μL of PB and the pH was adjusted.
3. The breast tumors homogenates (benign breast tumor (Fibroadenoma)) were prepared as described in section (2.1.7), except using PB-buffer instead of PBS at the same pH and concentration carried out the homogenization.

Procedure

1. One hundered microliters of each group homogenate (benign breast tumors (Fibroadenoma) and pre-post

menopausal malignant breast tumors (IDC)) containing (100,100 and 200 $\mu g.mL^{-1}$ protein) were incubated with 50 μL of ^{125}I -anti CA 15-3 antibody concentration (0.175,0.175 and 0.140 $gm.mL^{-1}$). The volume was made up to 500 μL with PB pH (7.0, 7.6 and 7.8) containing 0.01 M of the following halides: NaF, NaCl, NaBr and NaI in each assay tube. (A sample without the addition of any salt was used as a control).

2. The assay tubes were then incubated for 90min. at 45, 15 and 45°C for the three groups individually.

3. Steps 3, 4, 5 and 6 of the experiment (2.2.4.1) were repeated.

Calculations

1. The values of (B/T) % were calculated as described in section (2.2.4.1).

2. (B/T)% was plotted against halides concentrations.

2.2.4.6. The Effect of Monovalent and Divalent Cations on the Binding

Reagents

1. PB was prepared as described in section (2.1.6) without addition of NaCl.

2. Monovalent and divalent cations salts were prepared in concentration of (0.025 M) PB at pH (7.0,7.6 and 7.8) individually, by dissolving each of 0.0931gm of KCl, 0.0668gm of NH_4Cl, 0.2541 gm of $MgCl_2.6H_2O$, 0.1388 gm of $CaCl_2.2H_2O$, 0.2474gm of $MnCl_2.4H_2O$, 0.3150 gm of $CuSO_4.5H_2O$ and 0.1703gm of $ZnCl_2$, in a final volume 50 mL of PB and the pH was adjusted.

Procedure

1. The same steps mentioned in section (2.2.4.5) were followed to determinate the effect of monovalent and divalent of CA 15-3 in the tissues homogenates of (benign breast tumors (Fibroadenoma) and pre-and postmenopausal malignant breast tumors (IDC)) with [125]I -anti CA 15-3 antibody, except the PB buffer containing (0.025M) of the following salts: KCl, $NH_4.Cl$, $MgCl_2.6H_2O$, $CaCl_2.2H_2O$, $MnCl_2.4H_2O$, $CuSO_4.5H_2O$, $ZnCl_2$.

2. A sample without the addition of any salt was used as control.

Calculations

1. The values of (B/T)% were calculated as described in section (2.2.4.1)
2. (B/T)%was plotted against monovalent and divalent cations salts concentrations.

2.2.4.7. Recovery of CA 15-3

Reagents

1. All reagents are described previously in section (2.1.6) and (2.2.3).
2. Standard concentration of CA 15-3 200 $U.mL^{-1}$ was used.

Procedure

Known concentration of CA15-3 (200 $U.mL^{-1}$) was added to the three groups of tissues homogenates (benign breast tumors (Fibroadenoma), and pre-and post-menopausal malignant breast tumors (IDC)). The experiment was carried out at optimum conditions that were obtained in experiments of (2.2.4). The CA15-3 was determined according to the experiment in section (2.2.3).

Calculations

1. The bound (c.p.m) of the reaction mixture (standard CA 15-3 was added to tissue homogenate) with ^{125}I -antiCA15-3 antibody, represent the measured value.

2. The bound (c.p.m.) of CA 15-3 in tissue homogenate with ^{125}I –antibody CA 15-3 antibody only , represent the expected value.

3. The recovery % (yield%) was calculated as follows:

$$\text{Recovery}\% = \frac{\text{Measured values}}{\text{Expected values}} \times 100$$

Human breast tissues in this study were classified according to type of breast tumors (benign and malignant) and the malignant breast tumors were again classified into sub groups (premenopausal and postmenopausal). Each type was examined histologically according to WHO classification. Homogenization was carried out in a cold medium (i.e.4°C) to avoid protein denaturation [148,149], by proteolytic enzymes [150]. The filtration of the tissue homogenate through several layers of nylon gauze was used to remove any suspended pieces unhomogenized fragments and blood vessels.

Determination of CA 15-3 Levels in Sera of Breast Tumors Patients

CA 15-3 levels in sera of patients with benign breast tumors (group I) and pre and post-menopausal malignant breast tumors (group II and group III) were measured by IRMA method. These three groups were matched with a group of control subjects.

Table (2.5) summarizes the groups and the mean concentrations of CA15-3 for the control women and patients with benign breast tumors and pre-and post-menopausal malignant breast tumors.

Table (2.5) showes that CA15-3 levels in three different groups (benign breast tumors and pre-and post-menopausal malignant breast tumors) were significantly elevated ($p<0.05$) for benign breast tumors and highly significantly elevation ($p<0.0001$) for pre-and post menopausal malignant breast tumors respectively, as compared with the control.

The mean serum CA15-3 level of the control was found to be (17.26 ± 4.06 U.mL^{-1}) as shown in table (2-5), and the cutoff values was (25 U.mL^{-1}) that obtained from (mean +2 SD). This cutoff value is in agreement with Geraghty J.G[151], other study obtained that cutoff value of 40 U.mL^{-1} [152], 22 U.mL^{-1} [153], 30 U.mL^{-1} [154].

It has shown that widely different cutoff value which was described ranging from 20-40 U.mL^{-1} in different reference [155-159].

According to Bon et al [160] the upper limit of CA 15-3 of normal may be method-dependent. No association between the CA 15-3 and either age or menopausal status was found in the control group. Therefore , the cutoff values do not require adjustments related to these variables. These results were in agreements with Gion M.et.al. [68]. Figure (2.3) shows the distribution of the individual values of CA15-3 in sera of patients with benign breast tumors and pre-and post menopausal malignant breast tumors and control, were determined by using the standard curve in figure (2-2).

It was found that the mean of serum CA 15-3 concentration in 20 patients with benign breast tumors was 21.9 ± 6.6 U.mL^{-1} (mean ± SD). These results are in agreement with Hayes D.F. et.al [161]. The results show there was highly significant correlation between serum connections of CA15-3 in both groups pre-and post-menopausal status with control, while it was significantly lower in benign breast tumors status.

This is in agreement with Ichihara S. et.al [162]. Therefore all of the cases used in the binding studies were concentrated to this type of carcinoma (IDC) and this type is the common type of breast cancer. In Iraq very high levels of CA15-3 advanced disease and the value 5 to 10 times of normal suggest the presence of metastasis. Increasing numbers of metastatic sites correlate with increasing CA15-3 levels [163,164].

These findings suggest that higher levels of CA15-3 represent the breast cancer extent and reflect the cell differentiation and aggressiveness of the tumor. Therefore, it could be concluded that the determination of CA15-3 before surgical operative may be useful as a prognostic factor in breast cancer.

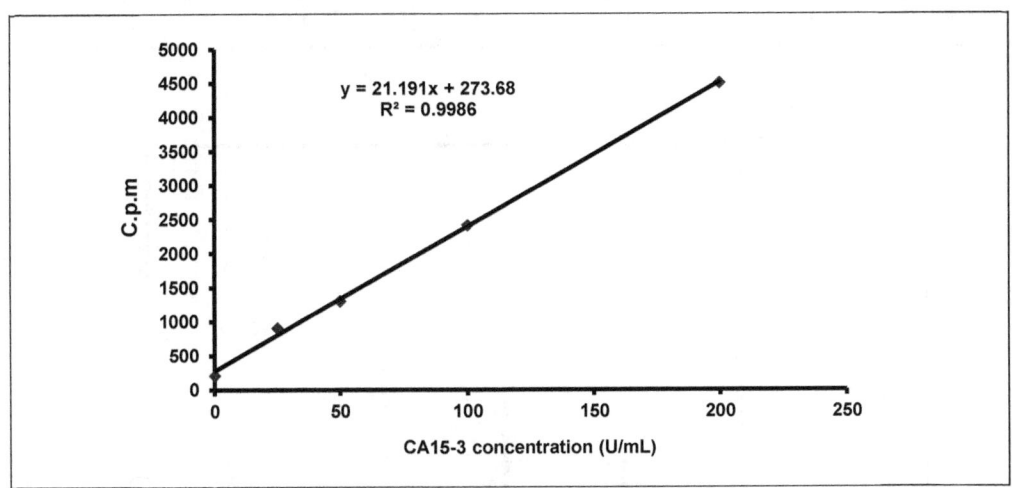

Figure (2.2): Standard curve of CA 15-3 determination in human sera by IRMA method.(All other details are explained in the text).

Table (2.5): Sera CA15-3 levels (U.mL⁻¹) in patients with benign and malignant breast tumors. (All details are explained in the text).

Group	Patients	No. of cases	Age range (Year)	Sera CA15-3 U.mL^{-1} (mean ± SD)	P values
I	Benign breast tumors	20	18-42	21.9 ± 6.6	P<0.05
II	Premenopausal malignant breast tumors	16	34-52	37.3 ± 6.8	P<0.0001
III	Postmenopausal malignant breast tumors	12	55-73	60.3 ± 10.9	P<0.0001
IV	Control	10	25-40	17.3 ± 4.06	--

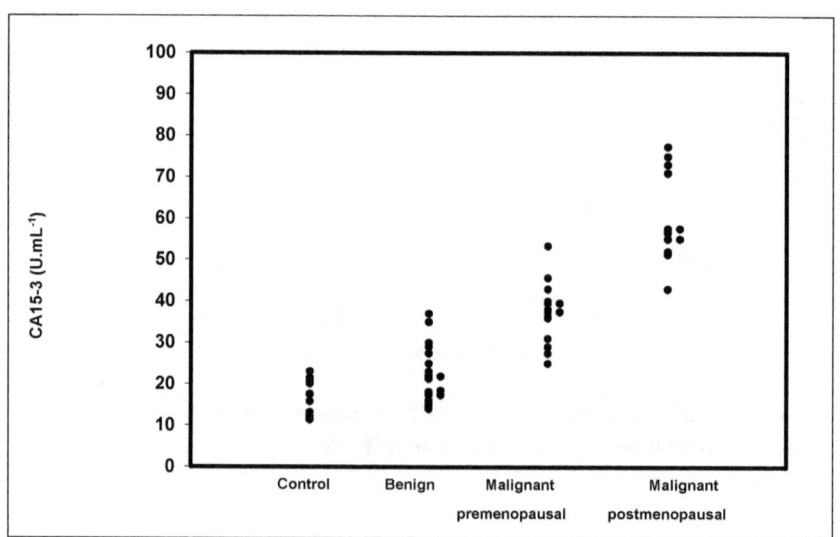

**Figure (2.3): Distribution of the individual
value of CA15-3 U.mL^{-1} in the
sera of benign and malignant**

breast tumors patients. (All other details are explained in the text).

Binding Studies of 125I-Anti CA15-3 Antibody with CA15-3 in Benign and Malignant Breast Tumors Homogenates

Preliminary Test of the Binding of 125I-Anti CA15-3 Antibody with CA 15-3 in Breast tumor homogenate

Supernatant and pellet formed at speed (4000 r.p.m.) in three groups of human breast tumor homogenate (benign breast tumors, pre-and post-menopausal malignant breast tumors) were used in this experiment. In each fraction CA 15-3 was detected through the incubation of ^{125}I-anti CA15-3 antibody with supernatant fraction and pellet individually for 2hrs at 37°C in PBS buffer as a medium to complete the reaction. The separation of the bound antibody from the unbound was carried out at 4000 r.p.m. for 45 min. to precipitate the (^{125}I-anti CA15-3 antibody/CA15-3) complex formed. Preliminary experimental conditions used in Table (2.6), which is show, the amount of binding B/T% in both fractions. The data revealed that CA15-3 was higher in incidence according to B/T%.

Table (2.6): Incidence of CA15-3 in supernatant and pellet fractions in three different breast homogenate.

Groups	(B/T)%		CA15-3 U.mL^{-1}
	Supernatant Fraction	Pellet Fraction	in supernatant fraction kit
Benign	6.20	3.40	90
Premenopausal	8.04	5.53	356
Postmenopausal	6.31	4.64	144

B/T% in supernatant is more than in pellet fractions of this speed (4000 r.p.m.). According to these results supernatant fractions was collected and the pellet was then discarded. The CA15-3 levels in the supernatant of breast tumors homogenate were determined according to IRMA method.

In general, results show that CA15-3 concentration in pre-and post-menopausal malignant breast tumors homogenates is more than benign breast tumors homogenates. These results are in agreement with the result obtained from B/T% from IRMA developed method.

From these results, it can be said that developed method was useful for determination CA15-3 in breast tumors homogenate using ^{125}I-anti CA15-3 antibody.

Factors Effecting of 125I-Anti CA15-3 Antibody Binding to CA15-3 in Breast Tumors Homogenates

The Effect of Different Amounts of Protein Concentration of the Tumor Homogenate on the Binding with 125I-Anti CA 15-3 Antibody

To obtain the optimum protein of homogenate for the binding of CA15-3 with ^{125}I-anti CA15-3 antibody, the supernatant homogenate containing increasing amount of

CA15-3 in the presence of fixed amount of ^{125}I-anti CA15-3 antibody as it was mentioned in section of (2.2.4.1).

Figure (2.4) represents the quantitative precipitation curve in which the amount of (^{125}I-anti CA15-3 antibody/CA15-3) complex in three groups (benign breast tumors and pre-and-post menopausal malignant breast tumors) was plotted as a function of CA15-3 concentration.

As shown in this figure, in the first phase of the reaction no precipitate was formed. The amount of precipitate increased until a point of maximum binding was reached. After this point as the amount of CA15-3 increased the amount of precipitate diminished; thus the increase in protein concentration which would increase the number of binding site and hence increase the percent of binding until the saturation state at (100, 100, and 200 μg.mL^{-1}) homogenate concentration for (benign breast tumors, pre-and post menopausal malignant breast tumors respectively).

The complex precipitate out of solution because of the multivalent nature of both molecules [165]. The radioactive antibody has two binding sites, it can cross-link antigenic sites

of two different CA15-3 molecules and can produce
maximum complex formation and therefore maximum
precipitate will occur. When CA15-3 is in greater excess,
large complex are again less probable.

In all subsequent experiments the amonts of (100, 100 and 200
μg.mL^{-1} protein) of tissue homogenate in benign breast tumors and pre-
and post menopausal malignant breast tumors were used according to the
result obtained in this experiment.

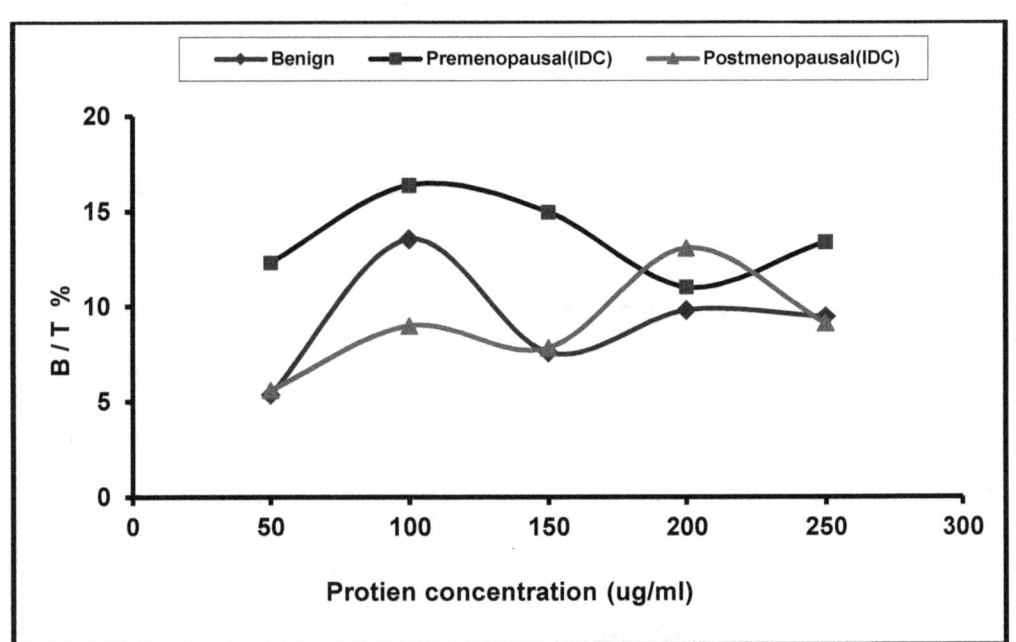

Figure (2.4): Influence of
increasing protein
concentration on the binding
with ^{125}I-anti CA15-3 antibody.
(All other details are explained
in the text).

The Effect of 125I-Anti CA15-3 Antibody Concentration on the Binding

The experiment was carried out in the presence of fixed amount of protein concentration of the homogenate and increasing concentration of ^{125}I-anti CA15-3 antibody.

The results are illustrated in figure (2.5). Which represent ^{125}I-anti CA15-3 antibody binding curve with supernatant fraction of benign breast tumor, pre-and post-menopausal malignant breast tumors. As shown in figure (2.5) it is obvious that the amount of (^{125}I-anti CA15-3 antibody/CA15-3) complex rises gradually, and then the breast tumor protein was saturated with ^{125}I-anti CA15-3 antibody. When the amount of antibody is in moderate excess, the probability of cross-linking of Ag by Ab in the incubation mixture is more likely, and hence large complex formation is favored. Then the maximum B/T percent was detected. The presence of (0.175, 0.175, 0.14 mg.mL^{-1}) of ^{125}I-anti CA15-3 antibody in benign, pre-and post-menopausal breast tumors homogenates give the optimum concentration of ^{125}I-anti CA15-3 antibody in three groups. Then the binding percent decreased as the amount of ^{125}I-anti CA15-3 antibody increased.

This is because all antigenic sites are covered with antibody and complex formation is inhibited [166]. These results indicate that the binding is principally dependent on the amount of the antibody in the reaction mixture [167].

According to the results of this experiment the above concentration of ^{125}I-anti CA15-3 antibody was used in the subsequent experiments.

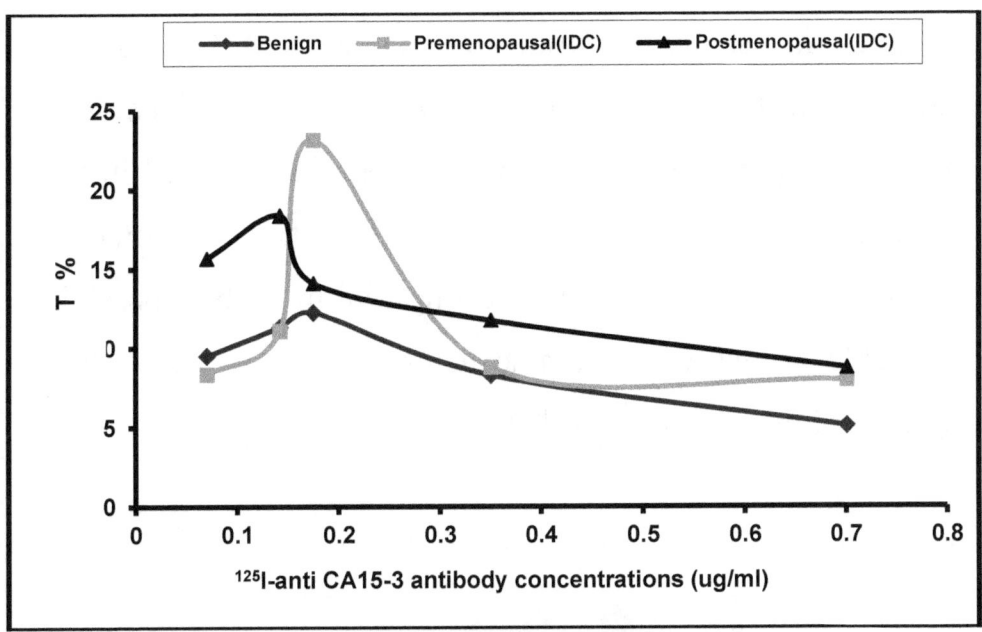

Figure (2.5): Effect of different concentrations of ^{125}I-anti CA15-3 antibody on the binding of with CA15-3. (All other details are explained in the text).

The Effect of pH on the Binding

Figure (2.6) shows the values of the binding of ^{125}I-anti CA 15-3 antibody to CA 15-3 in benign breast tumor, pre- and post-menopausal malignant breast tumors, at different pH values. Maximum value of the binding occurs at (pH 7, pH 7.6, pH 7.8) for benign breast tumor, pre-and post-menopausal malignant breast tumors respectively.

The formation of (^{125}I-anti CA 15-3 antibody/CA 15-3) complex is usually performed at pH between 6.8-8.0; the results indicate that the shift in the pH of the environment may affect the properties of CA 15-13 molecules involved in the binding. This effect may include the protonation deprotonation processes occurring within the possible ionizable groups of the amino acids present in the binding domain of these molecules [168].

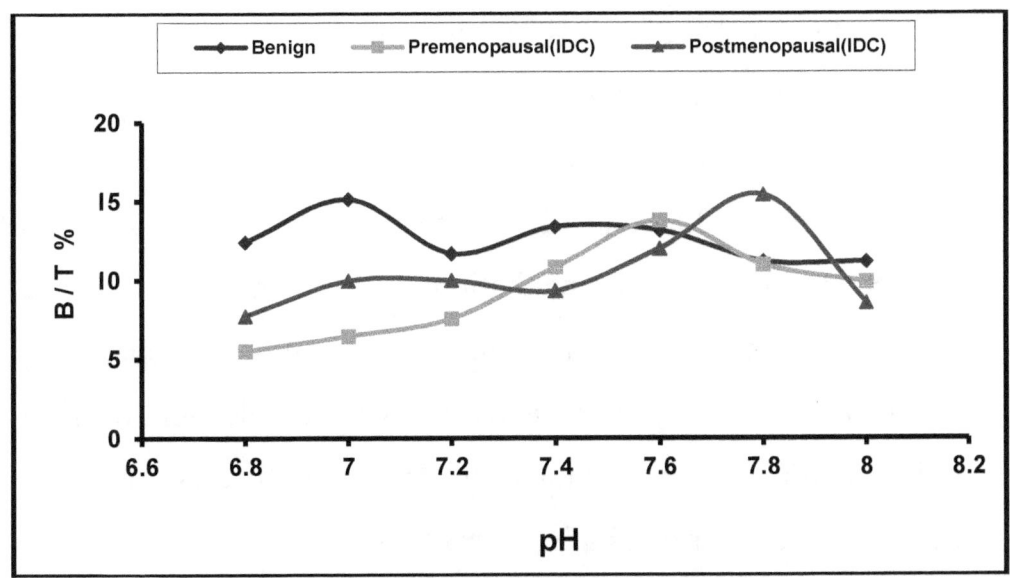

Figure (2.6): Effect of pH on the binding of ^{125}I-anti CA 15-3 antibody with CA 15-3 in breast tumors homogenates. (All other details are explained in the text).

Time Course of the Binding of 125I -Anti CA15-3 Antibody to CA 15-3 in Breast Tumors Homogenates

The results of time course pattern at different temperatures (5, 15, 25, 37, 45°C) indicate the ^{125}I-anti CA 15-3 antibody binding to crude fractions of CA 15-3 is temperature and time dependent process, as shown in figures (2.7), (2.8) and (2.9). The maximum binding was obtained at 45°C after incubation for 90 min. in crude fractions of benign breast tumors and postmenopausal malignant breast tumors respectively, whereas the binding in crude fractions of premenopausal malignant breast tumors occurs at 15°C after incubation for 30 min.

The decrease of the binding activity may be due to reversible dissociation of (^{125}I-anti CA 15-3 antibody/CA 15-3) complex after reaching the equilibrium state.

At 45°C the CA 15-3 molecule preserve the nature of protein structure and gave the maximum binding, but at higher temperature than 45°C denaturation may occur.

In the premenopausal malignant breast tumors the maximum binding occurs at 15°C for 30 min., in this state the energy is enough to overcome the energy barrier and give the maximum binding [(169)], the decrease in binding after 15°C may be due to proteolytic enzyme.

The difference in incubation time to give the maximum binding may be due to the different source of CA 15-3. According to this results, the binding studies of the subsequent experiments were carried out a 45°C for 90 min incubation for benign and postmenopausal breast tumors homogenate, whereas 15°C for 30 min. incubation for premenopausal malignant breast tumors homogenate.

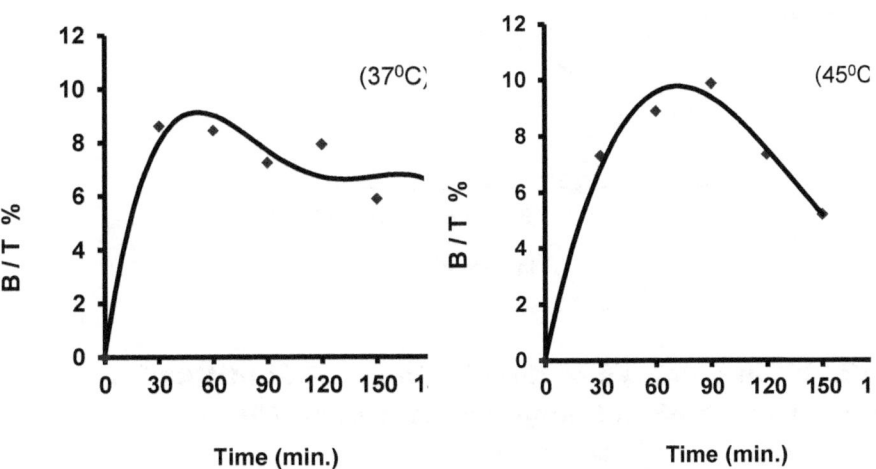

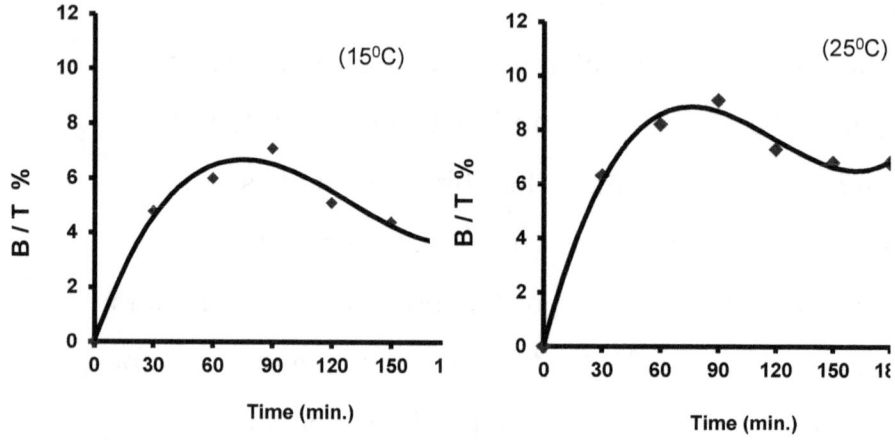

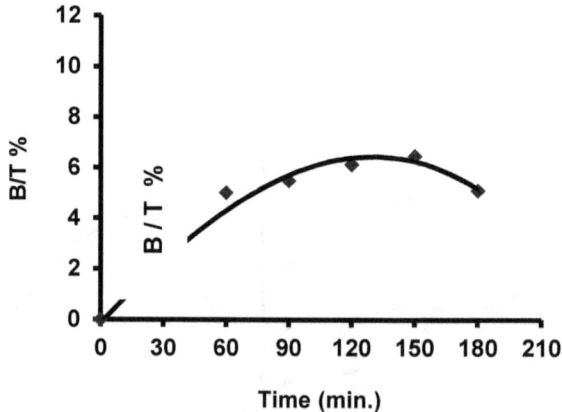

Figure (2.7): Time course of the binding of ^{125}I-antiCA15-3 antibody with CA15-3 in benign breast tumor. (All other details are explained in the text).

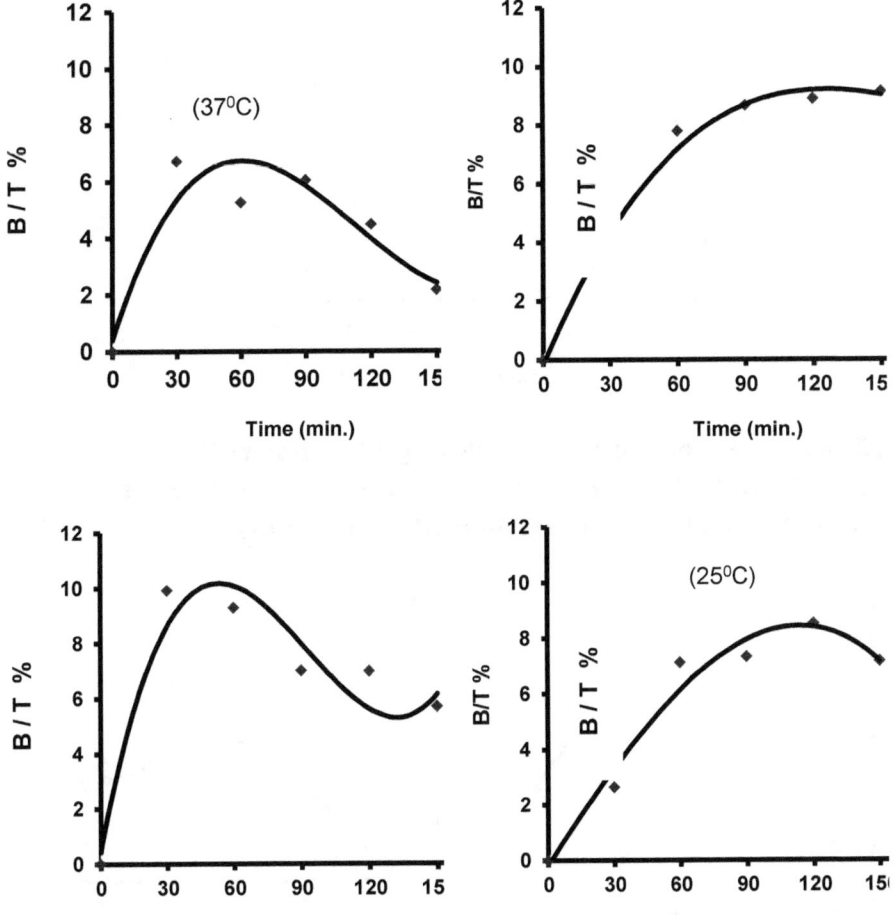

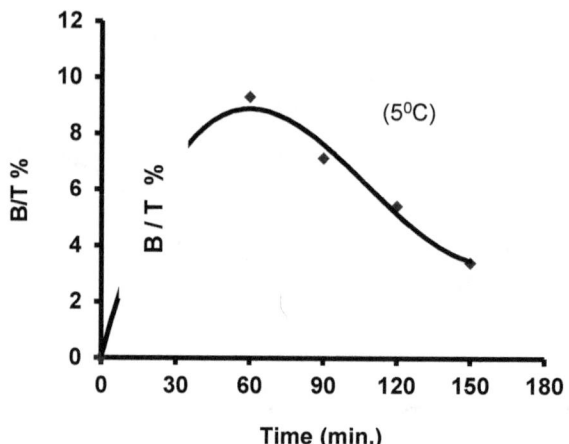

Figure (2.8): Time course of the binding of ^{125}I-antiCA15-3 antibody with CA15-3 in premenopausal malignant breast tumor. (All other details are explained in the text).

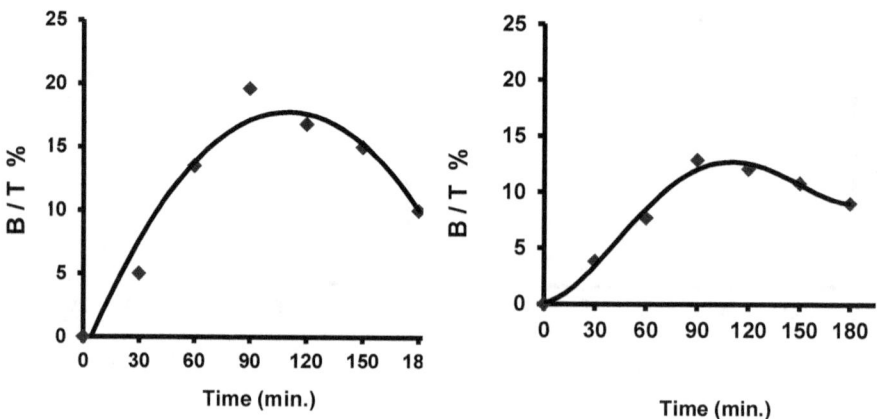

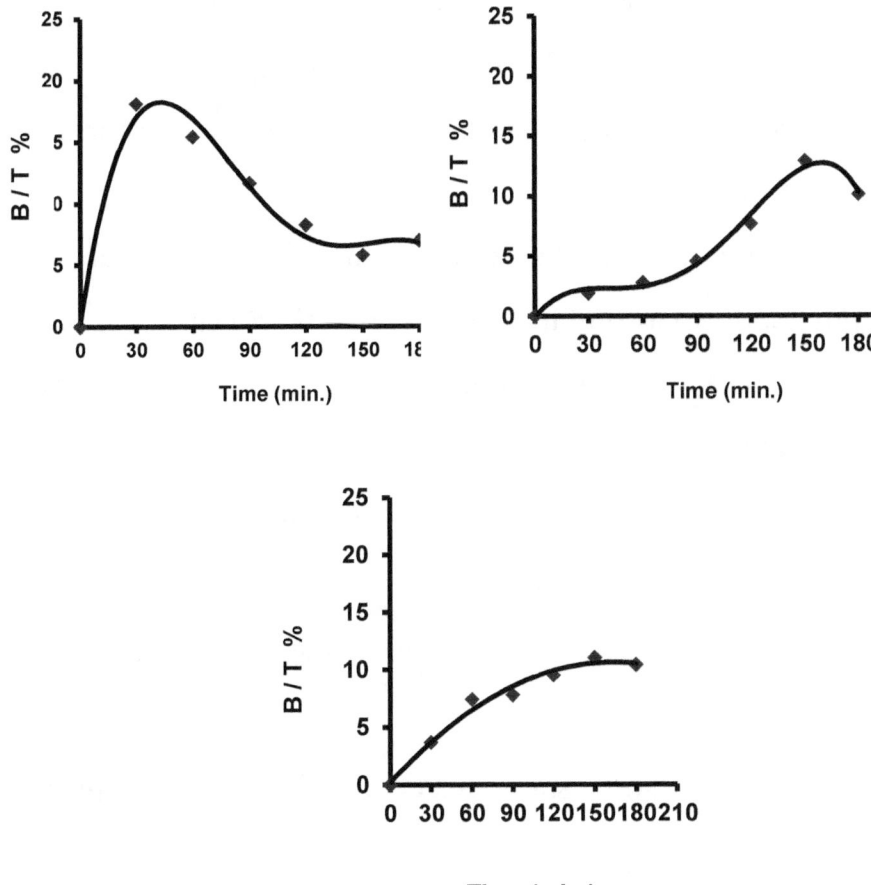

Figure (2.9): Time course of the binding of [125]**I-antiCA15-3 antibody with CA15-3 in postmenopausal malignant breast tumor. (All other details are explained in the text).**

The Effect of Different Halides on the Binding

Different sodium halides at 0.01 M concentration were investigated to study their action on the binding ^{125}I-anti CA 15-3 antibody with CA 15-3 in the three groups (benign breast tumors, Pre-and postmenopausal malignant breast tumors), as shown in figure (2.10).

The presence of the sodium halides in the incubation medium tends to promote the binding of ^{125}I-anti CA 15-3 antibody to CA 15-3 in these groups, the following sequence of effects have occurred.

1.Benign breast tumor tissue homogenate

NaI > NaBr > NaCl > NaF

2.Premenopausal breast cancer tissue homogenate

NaCl > NaI > NaBr > NaF

3.Postostmenopausal breast cancer tissue homogenate

NaCl > NaBr > NaF > NaI

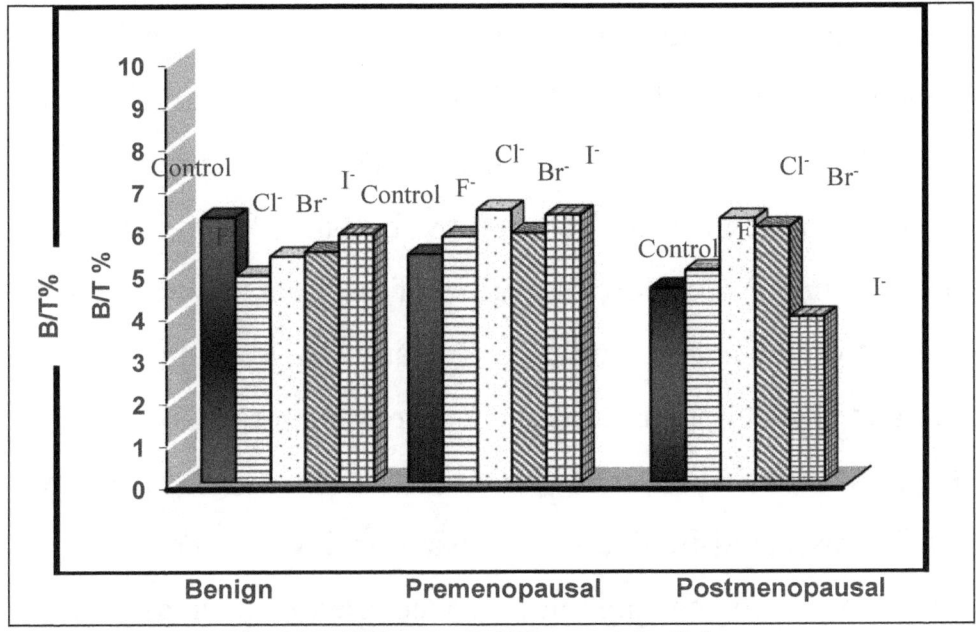

Figure (2.10): Effect of different halides on the binding of ^{125}I-anti CA 15-3 antibody with CA 15-3. (All other details are explained in the test).

As shown in figure (2.10), the sodium halides inhibited the binding in benign breast tumors, according to the decreasing ionic radius and increasing radius of hydration. It seemed that fluoride ion causes lower binding, this could be due to higher electro negativity of fluoride ion that tend to interact with the positive residue in the binding site of the antibody and/or the antigen which lead to decrease the interaction between CA 15-3 and its antibody [170].

Melander and Horvath (1977) reported that the effect of halide salt type on hydrophobic interactions is quantified by its molar surface tension increment (MSTI) that is a measure of the increase in surface tension by the salt [171]. On the other hand, figure (2.10) shows the effect of different halides salts at 0.01 M on the extent binding of ^{125}I-anti CA 15-3 antibody to pre-and postmenopausal malignant breast tumors homogenate. It seems that halides salts increased the binding, especially NaCl, this could be due to that NaCl in lower concentration (0.15M) or in physiological concentration, increased the binding between CA 15-3 and its antibody [172].

The Effect of Monovalent and Divalent Cations on the Binding

The effect of different salts on the extent of binding of ^{125}I-anti CA 15-3 antibody to CA 15-3 in benign and malignant breast tumors are shown in figure (2.11).

The results indicate that the binding process is sensitive to the presence of cation metal ions. $CuSO_4.5H_2O$ at concentration (25mM) was shown to increase the binding more than other divalent cations, while $ZnCl_2$ increased the binding less than other divalent cations. One hypothesis assumes that salts may alter the nature of the hydrophobic forces controlling stabilization of the complex formed and these vary depending on the nature of the interacting groups [172]. From the results illustrated in figure (2.11), it is suggested that these salts maybe provide some conformational changes in the CA 15-3 and the charged groups of the binding domain of the antibody and antigen molecule [173,174], that hinder maximal binding are shielded. If the interaction is dominated by ionic strength, high salt concentration lowers the affinity.

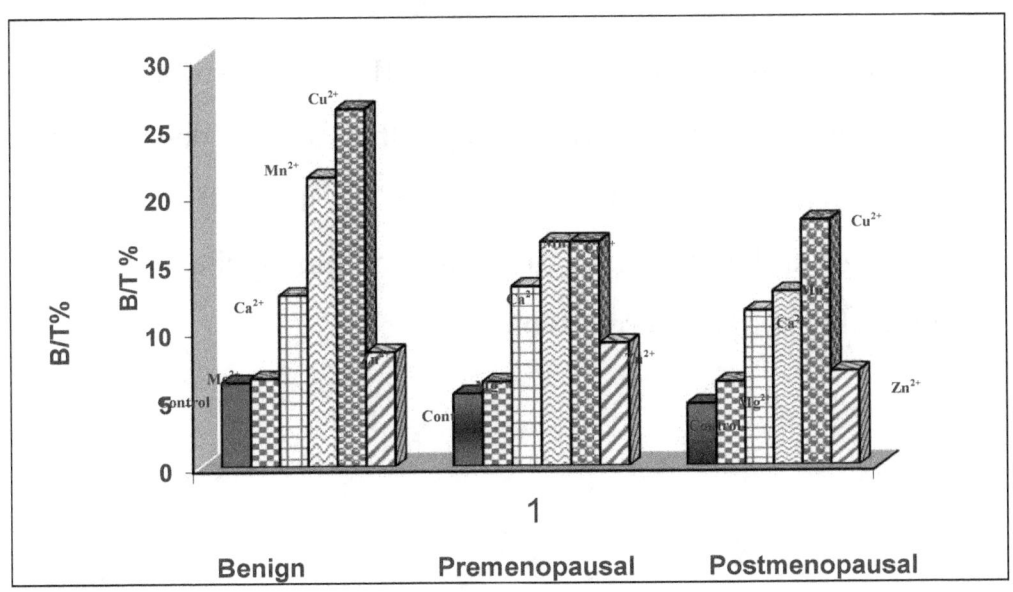

Figure (2.11): Effect of different divalent cations on the binding of ^{125}I-anti CA 15-3 antibody with CA 15-3. (All other details are explained in the text)

Figure (2.12) shows the effect of monovalent cations (KCl and NH_4Cl) on the extent of the binding of CA 15-3 to its antibody ^{125}I-anti CA 15-3 in benign and malignant breast tumors. KCl at 25mM was shown to increase the binding in benign and premenopausal malignant breast tumors as compared with the control value, while KCl at the same concentration slightly inhibiting the binding in postmenopausal malignant breast tumors. These results may be due to conformational changes. NH_4Cl at 25 mM was shown to inhibit the binding but to a lesser extent.

This result shows that NH_4Cl effect on the binding is nearly unremarkable. Presumably, the lesser degree of hydration permits greater interaction of the salt with an anionic group located in the antibody-combining site and then inhibits the complex formation.

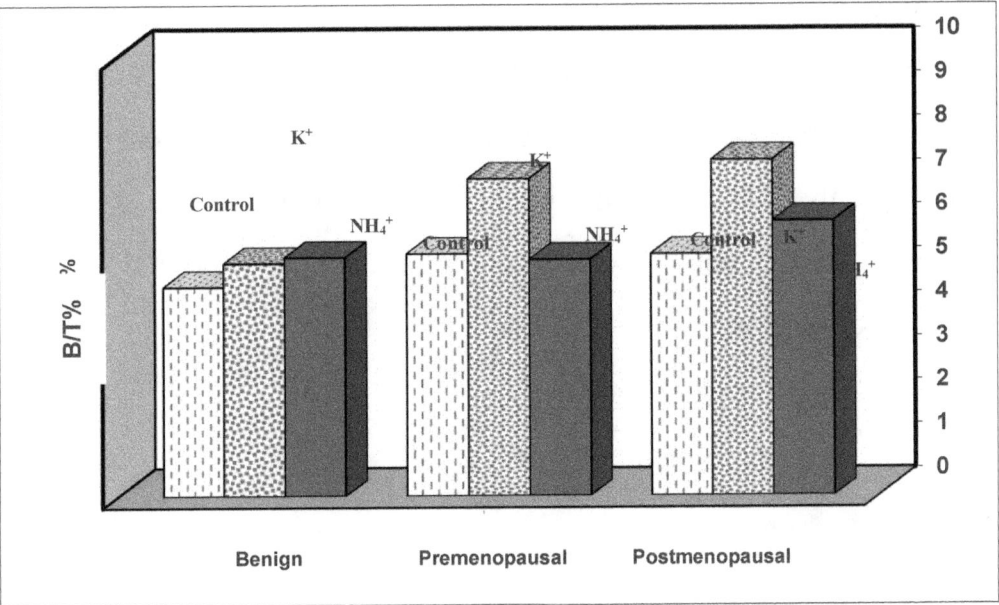

Figure (2.12): Effect of different monovalent cations on the binding of ^{125}I-anti CA 15-3 antibody with CA 15-3. (All details are explained in the text).

Recovery of CA 15-3

This method was used to estimate the percent recovery of CA 15-3 in supernatant fractions of benign and malignant breast tumors homogenates. The results are summarized in table (2.7) and indicate that the CA 15-3 extracted from benign breast tumors, and CA15-3 extracted from malignant breast tissues homogenate were recovered more than CA 15-3 extracted from postmenopausal malignant breast tumors homogenates were recovered more than CA 15-3 extracted from premenopausal malignant breast tumors homogenates. Also the results indicate that total CA 15-3 can be determined through the developed method of immunoradiometric assay, as well as the percent of recovery indicates the precision of the used method.

Table (2.7): Recovery of CA 15-3. (All other details are explained in the text).

Type of CA 15-3	Measured B/T	Expected B/T	Recovery% Measured / Expected
Benign (Fibroadenoma)	16.23	24.84	65.34
Premenopausal (IDC)	20.04	27.64	72.50
Postmenopausal (IDC)	24.20	26.80	90.30

s wit**Abstract**

Gel filtration chromatography technique was used for partial purification of CA15-3 from breast tumor homogenates.

The results revealed the presence of one form of CA15-3 with a high molecular weight (440 KD). This type possesses a high affinity for the binding to its antibody ^{125}I-anti CA15-3 at the same conditions performed in section (2.2.4) of chapter two.

The elution volume (V_e) and the K_{av} value for elution of CA15-3 from sepharose CL-4B column were calculated. The experiments of optimum conditions of the binding between the partially purified CA15-3 and ^{125}I-anti CA15-3 antibody were determined, in benign breast tumor and premenopausal malignant breast cancer homogenates.

Studies on the stability of both partial purified CA15-3 and crude CA15-3, show that the crude CA15-3 was more stable than the purified CA15-3.

Chapter Three

Purification of CA-15-3

CA15-3 is high molecular weight glycoprotein (>400 KD) identified at the apical side of alevoli and duct of mamary glands [83]. Several authors have isolated, purified and characterized CA15-3 from different sources;either by the isolation of CA15-3 from a breast cancer patient's sera, using affinity chromatography, gel filtration, and then characterized by SDS-PAGE[175,176] ,or by purifing a high molecular weight glycoprotein from human milk and breast carcinoma by using gel filtration, affinity chromatography and then PAGE [177]. In the present study , benign breast tumors and premenopausal malignant breast cancer were used as a source for partial purification of CA15-3 and then determination its yield. The factors effect the binding of partial purified CA15-3 to its antibody ^{125}I-anti CA15-3 antibody were also studied.

Materials and Methods

3.1. Materials

3.1.1. Chemicals

All chemical and reagents mentioned in section (2.1.1) and (2.1.6) were used in the experiments of this chapter.

3.1.2. Instruments

All instruments mentioned in section (2.1.2) were also used in the experiments of this chapter.

3.1.3. Patients

The same patients tissues mentioned in section (2.1.5) were used in the following experiments. Benign breast tumor and premenopausal malignant breast cancer homogenates that showed maximal binding in the preliminary test in section (2.2.3) were used for the purification of CA15-3.

3.2. Methods

3.2.1. Isolation of CA15-3 by Sepharose CL-4B Column

3.2.1.1. Preparation of the Column

The dimensions of the column were chosen according to the following equation [150].

$$\text{Diameter} = \sqrt{\frac{m}{10}}$$

Where:

m= amount of protein in mg.

L = 30 x diameter

Where:

L : length of the column

3.2.1.2. Preparation of Phosphate Buffered Saline

PBS buffer pH 7.0 containing 0.02% sodium azide was prepared as described previously in section (2.1.6).

3.2.1.3. Preparation of the Gel

The gel was prepared by allowing the pre-swollen gel to swell again in PBS buffer (0.05 M) pH 7.0, then left to settle and the excess of buffer was decanted. The step was repeated several times. Suction was then used to degas the gel and slurry was left for 24 hrs to equilibrate with buffer.

The swollen gel was suspended and carefully poured into a vertical glass-column (0.7 x 30 cm) down the wall using a glass rod. After the gel has settled, the column was equilibrated with PBS for 24 hrs.

3.2.1.4. Void Volume Determination

The void volume of the column was determined by using blue dextran 2000 at concentration of 2 mg.mL^{-1} dissolved in PBS buffer pH 7.0 , then the elution was carried out with the same buffer at a flow rate of 20 mL.hrs^{-1}.

Fractions of 2 mL were collected and their absorbance was measured at 600 nm. Figure (3-1) shows the elution profile of blue dextran 2000. The volume of the buffer required to elute the blue dextran which represents the void volume was (6 mL).

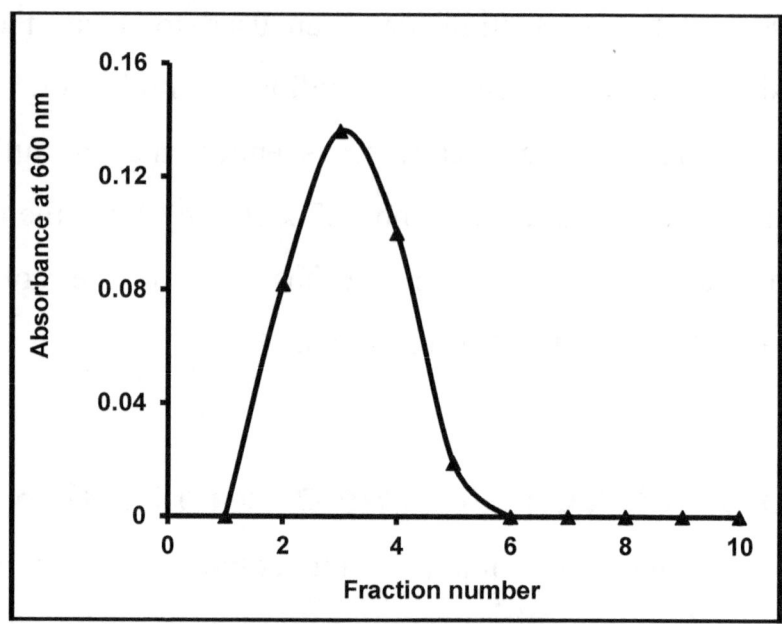

Figure (3.1): The elution profile of blue dextran 2000. (All other details are explained in the text).

3.2.1.5. Column-Calibration

The column was calibrated by gel filtration kit, purchased from pharmacia fine chemicals which contained standard proteins. Standard protein solutions were prepared according to the manufacturers instructions, then applied through two 0.5 mL portions, proteins 1 and 3 in the first portion, protien 2 and 4 in the second portion. Elution was carried out with PBS buffer at a flow rate of 20 ml.hrs^{-1}. the absorbance of the fractions collected was measured at 280 nm to evaluated the elution volume (Ve) of the standard protein.

Standard Proteins

Pharmacia calibration kit for determination of M.wt by gel filtration was used. The kit comprises the highly purified proteins and their high M.wt are detailed in table (3.1).

Protein	M.wt (KD)	Conc. mg.mL-1
Thyroglobulin	669	4.0
Ferritin	440	1.0
Catalase	322	6.0
Aldolase	158	6.0

Calculations

The K_{av} values of the proteins eluted were determined using the following equation:

$$K_{av} = \frac{V_e - V_o}{V_t - V_o}$$

Where:

V_o= Void volume

V_e=Elution volume of each protein

V_t=Total gel - bed volume.

The calibration curve of K_{av} values vs. log M.wt. of the proteins were plotted.

3.2.1.6. Separation Procedure

Reagents

PBS buffer pH (7.0, 7.2 and 7.6) containing 0.02% sodium azide was prepared as described previously in section (2.1.6).

Procedure

The sample of tissue homogenate (0.5 mL) containing approximately 3.43 mg protein was applied to the surface of gel , equilibrated with 0.15 M PBS buffer pH 7.2 for benign and premenopausal malignant breast tumor respectively. The sample was eluted by using the same buffer pH (7.0 and 7.6) for (benign and premenopausal malignant breast tumors respectively) with a flow rate of 20 mL.hrs^{-1} and fractions volume 2 mL were collected, gel filtration was carried out at 10 °C. The protein content of each fraction was determined using Lowry.et.al method [147].

The fractions contained CA15-3 were identified by the assay method. The binding of each fraction was calculated and plotted against the elution volume. The degree of purification (folds) of CA15-3 was calculated from the following formula.

$$\text{Purification fold of CA 15-3} = \frac{\text{Specific binding of purified CA 15-3}}{\text{Specific binding of crude CA 15-3}}$$

Then yield % was determined as follows:

$$\text{Yield \%} = \frac{\text{Total protein content of purified CA15-3}}{\text{Total protein content of crude CA15-3}} \times 100$$

3.2.1.7. Dialysis for Concentration

After preparing dialysis tube, the fractions that contained high levels of the binding activity were pooled and concentrated by dialyzing against sucrose at 4 °C for 2hrs to get the required concentration to be used in the next experiments.

3.3. The Choice of the Optimum Conditions for the Binding of the Partially Purified CA15-3 to 125I-Anti CA15-3 Antibody

3.3.1. Optimum Protein Concentration

Reagents PBS buffer pH 7.0 and 7.6 was prepared as described previously in section (2.1.6).

Procedure

One hundred microliters of increasing amount (50,100,150,200 and 250) $\mu g.mL^{-1}$ protein of the dialyzable fractions of the partially purified CA15-3 from benign breast tumor was incubated with 50 μL of ^{125}I-anti CA15-3 antibody (0.35 $mg.mL^{-1}$) and completed to a final volume of 500 μL with 0.15 M PBS pH 7.0. The assay tubes were incubated for 90 min. at 45 °C. Two additional tubes, containing 50 μL (0.35 $mg.mL^{-1}$) of ^{125}I-anti CA15-3 antibody only, for total radioactivity computation, were set a side until counting.

Steps 4,5 and 6 of the experiment (2.2.4.1) were repeated. The same experiment was repeated on premenopausal malignant breast tissues homogenates (100 $\mu g.mL^{-1}$ protein) with PBS buffer pH 7.6 and incubation time for 90 min at 15 °C.

Calculations

The (B/T) % was calculated as mentioned in experiment (2.2.3) and plotted against increasing amounts of protein concentration.

3.3.2. Influence of 125I-Anti CA15-3 Antibody on the Binding

Reagents

PBS buffer pH 7.0 and 7.6 was prepared as described previously in section (2.1.6).

Procedure

Fifty microliters of increasing concentration (0.070, 0.140, 0.175, 0.210, 0.245, 0.280 mg.mL^{-1}) of ^{125}I-anti CA15-3 antibody were added to 100 μL (150 μg.mL^{-1} protein) of partially purified CA15-3 from benign breast tumors. The reaction was completed to 500 μL with PBS pH 7.0. The assay tubes were incubated for 90 min at 45 °C. Two additional tubes containing increased concentration of ^{125}I-anti CA15-3 antibody only, for total counts were counted. Steps 4,5 and 6 of the experiment (2.2.4.1) were repeated. The same experiment was repeated on premenopausal malignant breast tissues homogenate (100 μg.mL^{-1} protein) with PBS pH 7.6 and incubation time for 90 min at 15 °C.

Calculations

The (B/T) % was calculated as mentioned in experiment (2.2.3) and plotted against increasing concentration of ^{125}I-anti CA15-3 antibody.

3.3.3. Optimum pH

Reagents

PBS buffer pH (6.8, 7.0, 7.2, 7.4, 7.6, 7.8, and 8.0) was prepared as described previously in section (2.1.6).

Procedure

To determine the optimum pH, 100 µL of a dialyzable fractions of partially purified CA15-3 from benign breast tumors (150 µg.mL^{-1} protein) were added to 20 µL of ^{125}I-anti CA15-3 antibody (0.140 mg.mL^{-1}). The volume of each fraction was completed to 500 µL with 0.15 M PBS of different pH (6.8 , 7.0 ,7.2 , 7.4 , 7.6 , 7.8 , 8.0). The assay tubes were incubated for 90 min at 45 °C. Two additional tubes, containing 20 µL (0.140 mg.mL^{-1}) of ^{125}I-anti CA15-3 antibody only , for total count , were set aside until counting. Steps 4,5 and 6 of experiment (2.2.4.1) were repeated. The same experiment was repeated on premenopausal malignant breast tissues homogenates (100 μg,mL^{-1} protein) and 25 μL (0.175 mg.mL^{-1}) of ^{125}I-anti CA15-3 antibody was incubated for 90 min at 15 °C.

Calculations

The (B/T) % was calculated as mentioned in experiment (2.2.3) and plotted against their corresponding pH values.

3.3.4. Optimum Temperature

Reagents

PBS buffer pH 7.0 was prepared as described previously in section (2.1.6).

Procedure

Twenty microliters (0.140 mg.mL^{-1}) of ^{125}I-anti CA15-3 antibody was added to 100 μL dialyzable fractions of partially purified CA15-3 from benign breast tumors (150 μg.mL^{-1} protein).

The volume of reaction was completed to 500 μL with 0.15 M PBS buffer pH 7.0. The assay tubes were incubated for 90 min at 45 °C. The same steps were repeated at (37, 25,

15, 5°C). Two additional tubes containing 20 μL (0.140 mg.mL^{-1}) of ^{125}I-anti CA15-3 antibody only, for total count, were set aside until counting. Steps 4,5 and 6 of experiment (2.2.4.1) were repeated.

The same experiment was repeated on the premenopausal malignant breast tissues homogenates (100 μg.mL^{-1} protein) and 25 μL (0.175 mg.mL^{-1}) of ^{125}I-anti CA15-3 antibody in PBS buffer pH 7.0, with incubation time 90 min at 15 °C. The experiment was repeated at different temperatures (45, 37, 25 and 5 °C).

Calculations

The (B/T) % was calculated as mentioned in experiment (2.2.3) and plotted versus temperatures of incubation.

3.3.5. The Effect of Incubation Time

Reagents

PBS buffer pH 7.0 was prepared as described previously in section (2.1.6).

Procedure

Twenty microliters (0.140 mg.mL^{-1}) of ^{125}I-anti CA15-3 antibody were added to 100 μL of dialyzable fractions of partially purified CA15-3 from benign breast tumors containing (150 μg.mL^{-1} protein). The reaction volume was completed to 500 μL with 0.15 M PBS buffer pH 7.0 , then incubated

at 37 °C for (30, 60, 90, 120, 150, 180 min). Two additional tubes counting 20 μL (0.140 mg.mL^{-1}) of ^{125}I-anti CA15-3 antibody for total counts , were set aside until counting. Steps 4,5 and 6 of the experiment (2.2.4.1) were repeated. The same experiment was repeated on the premenopausal malignant breast tissues homogenates (100 μg.mL^{-1} protein) and 25 μL (0.175 mg.mL^{-1}) of ^{125}I-anti CA15-3 antibody with 0.15 M PBS buffer pH 7.0 and incubated at 15 °C for (30, 60, 90, 120, 150 and 180 min).

Calculations

The (B/T) % was calculated as metioned in experiment (2.2.3) and plotted versus the time of incubation for each group.

3.3.6. Stability of CA15-3 at −20 oC

Reagents

PBS buffer pH 7.0 was prepared as described previously in section (2.1.6).

Procedure

Crude and purified CA15-3 were stored at −20 °C for several time intervals. The frozen specimen was thawed at the end of each interval and the binding activity was measured at optimum conditions as described in section (2.2.4) and (3.6). The remaining activity was calculated and plotted against storage periods.

Calculations

The (B/T) % was calculated as mentioned in experiment (2.2.11) and plotted versus time storage for each group.

3.4. Results and Discussion

Partial Purification of CA15-3

Isolation of CA15-3 was performed by gel exclusion chromatography technique. CA15-3 was found to be separated from aggregates and other protein having smaller molecular weight by sepharose CL-4B. Figure (3-2, A & B) shows the elution profile of CA15-3 from benign breast tumors and premenopausal malignant breast cancer homogenates. The homogenate was loaded on the column as described in section (3.2.1). The void volume (V_o) of column was (6 mL) as predicted from the elution profile of the blue dextran. The elution was performed with PBS buffer. The resultant fractions containing the binding activity of CA15-3 were collected, pooled and concentrated, then subjected to protein determination as in section (2.2.1).

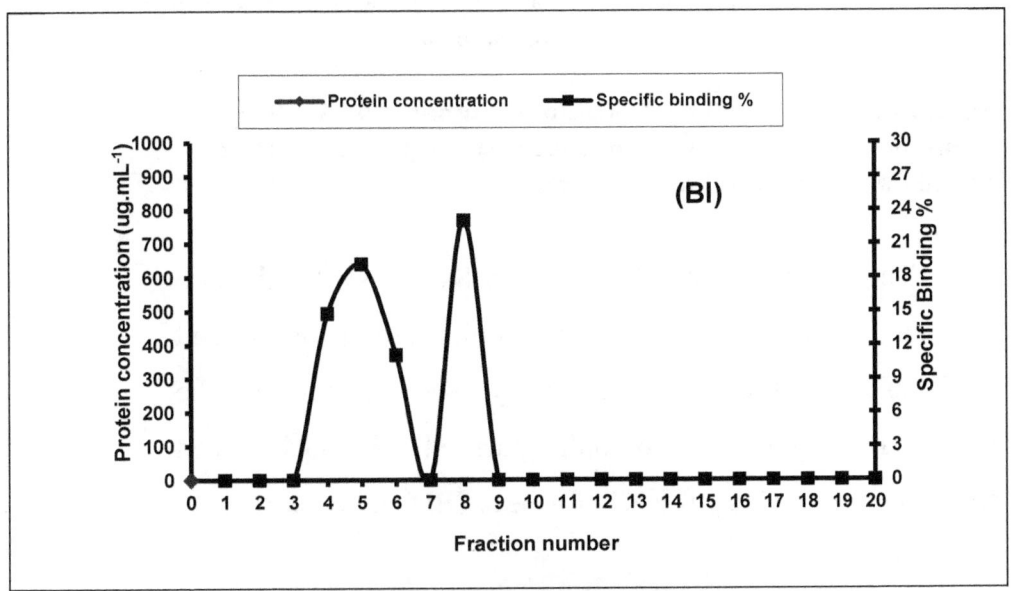

Figure (3.2A): The elution profile of human CA15-3 from benign breast tumors (BI). (All other details are explained in the text).

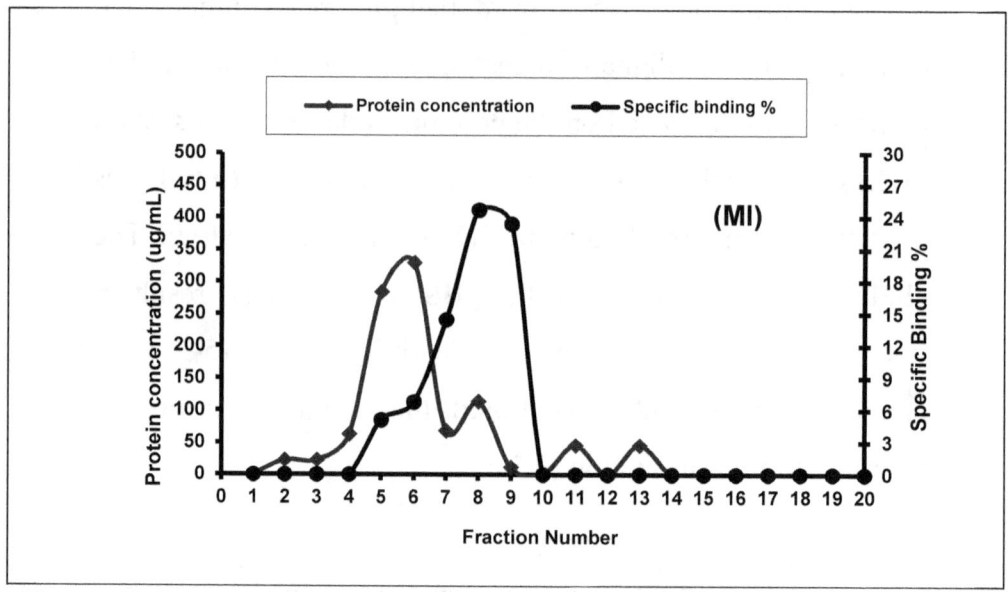

Figure (3.2B): The elution profile of human CA15-3 from premenopausal malignant breast cancer (MI). (All other details are explained in the text).

The elution volume V_e and then K_{av} values for the two peaks of CA15-3 (BI & MI) from benign breast tumors and malignant breast cancer respectively were calculated. The molecular weight of the partially purified CA15-3 obtained from figure (3-3) was 440 KD for peak (BI) and peak (MI) in two cases.

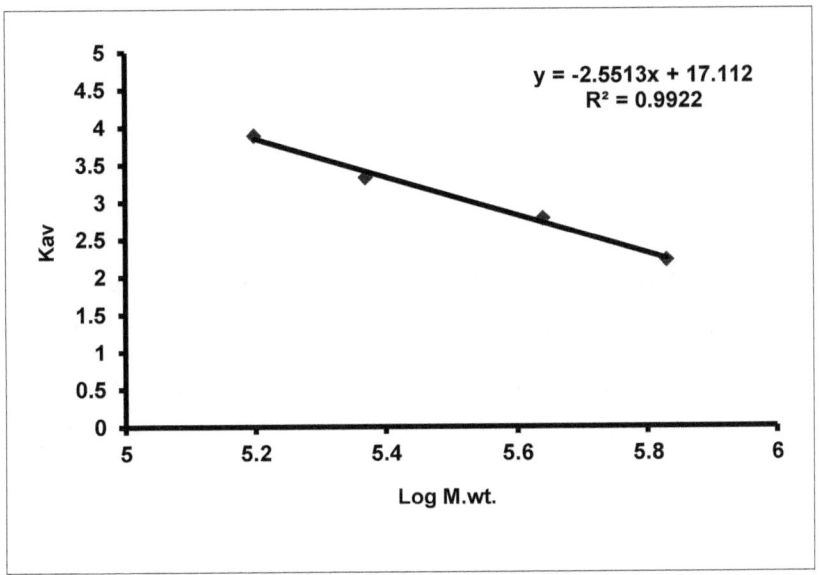

Figure (3.3): Calibration curve for determination of M.wt by gel filtration chromatography. (All other details are explained in the text).

The values ranged from 300-450 KD [178-180]. Peaks of partially purified CA15-3 may be heavily aggregated, CA15-3 was obtained near the void volume of the column under separation conditions. From these results it was concluded that these components are capable of binding to the ^{125}I-anti CA15-3 antibody with different affinities and in general CA15-3 type (BI) have lower binding affinities than CA15-3 type (MI), the isolation of CA15-3 from benign breast tumors on gel filtration column showed 3.02 folds of purification for peak (BI), while the isolation of CA15-3 from premenopausal malignant breast cancer showed 5.0 folds of purification. Table (3-2) illustrates the purification parameters for the different purified CA15-3 forms isolated by gel exclusion chromatography technique. The glycosylation of the protein backbone may differ in carcinoma cells from normal epithelial cells causing a wide range of molecular weight for this mucin [181].

Table (3.2): Partial purification of CA15-3 by gel filtration. (All other details are explained in the text).

	CA15-3 Source	Total protein mg.mL^{-1}	Specifically bound ^{125}I-anti CA15-3	Specifically binding ^{125}I-anti CA15-3/mg protein	Yield %	Purification fold
Benign	Crude extract	3.43	10.17	2.97	100	1.00
Benign	Gel filtration on sepharose CL-4B	2.91	30.70	10.55	84.84	3.02
Malignant	Crude extract	3.43	8.18	2.39	100	1.00
Malignant	Gel filtration on sepharose CL-4B	2.21	40.93	18.52	64.43	5.00

The Choice of Optimum Conditions for the Binding of Partially Purified CA15-3 with 125I-Anti CA15-3 Antibody

Optimum Protein Concentration

Figure (3-4) shows the effect of increasing amounts of partially purified CA15-3 to a fixed amount of ^{125}I-anti CA15-3 antibody to produce (^{125}I-anti CA15-3 antibody/CA15-3) complex, that grow in size until they formed a precipitate. Above this zone an equivalence between CA15-3 and its antibody concentration is obtained, and amount of complex shows no further increases. A further addition of CA15-3 give rise to a solubilization of complex. The results revealed that 150 µg protein was the most appropriate concentration for the binding of (BI) and 100 µg protein for (MI). From these results, it could be concluded that the binding of ^{125}I-anti CA15-3 antibody with its partially purified CA15-3 (BI) needed a higher amount of protein concentration than partially purified CA15-3 (MI). This is may be due to lower concentration of CA15-3 in benign breast tumor as compared with malignant breast tumors. According to these results, in all subsequent experiments, (150 µg.mL^{-1} protein) in benign breast tumors and (100 μg.mL^{-1} protein) in malignant breast tumors were used, since they give the highest binding.

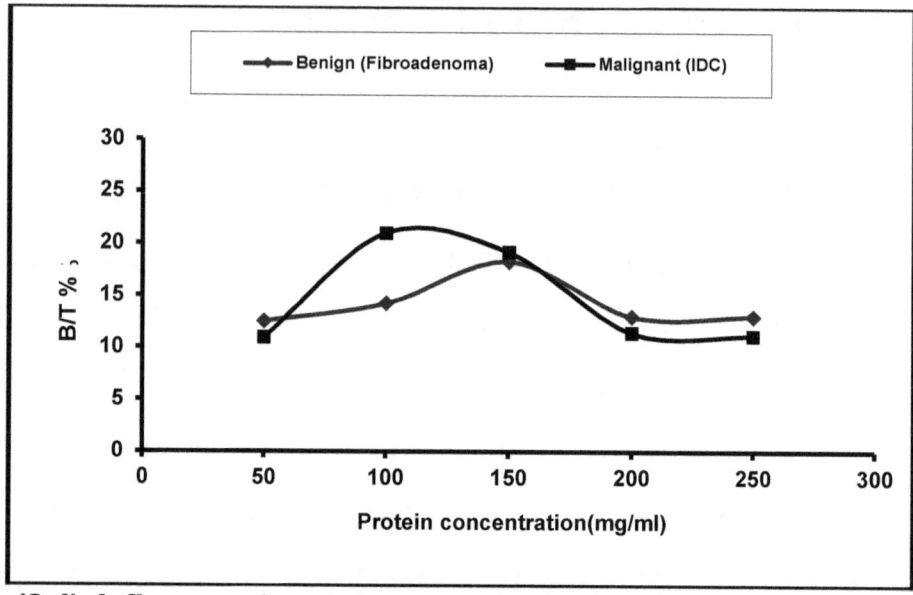

Figure (3.4): Influence of protein concentration on the binding of ^{125}I-anti CA15-3 antibody with partially purified CA15-3 from breast tumors. (All other details are explained in the text).

Influence of 125I-anti CA15-3 Antibody on the Binding

Figure (3.5) illustrate the effect of ^{125}I-anti CA15-3 antibody concentration on the binding with partial purified CA15-3 from benign breast tumors and premenopausal malignant breast cancer.

The maximum binding obtained when 0.140 mg.mL^{-1} of antibody in benign breast tumors and 0.175 mg.mL^{-1} of antibody in malignant breast tumors were used. From these results, it was found that (BI) purified fraction was saturated with small concentration of ^{125}I-anti CA15-3 antibody than those required for (MI). This is may be due to the increasement of the epitope (is the part of an antigen molecule that binds to any single antigen-combining site) [182] in partially purified CA15-3 in malignant breast tumors as compared to benign breast tumors.

According to these results, in all subsequent experiments (0.140 mg.mL^{-1}) and (0.175 mg.mL^{-1}) of ^{125}I-anti CA15-3 antibody in benign and malignant breast tumors were used, since they give the highest binding.

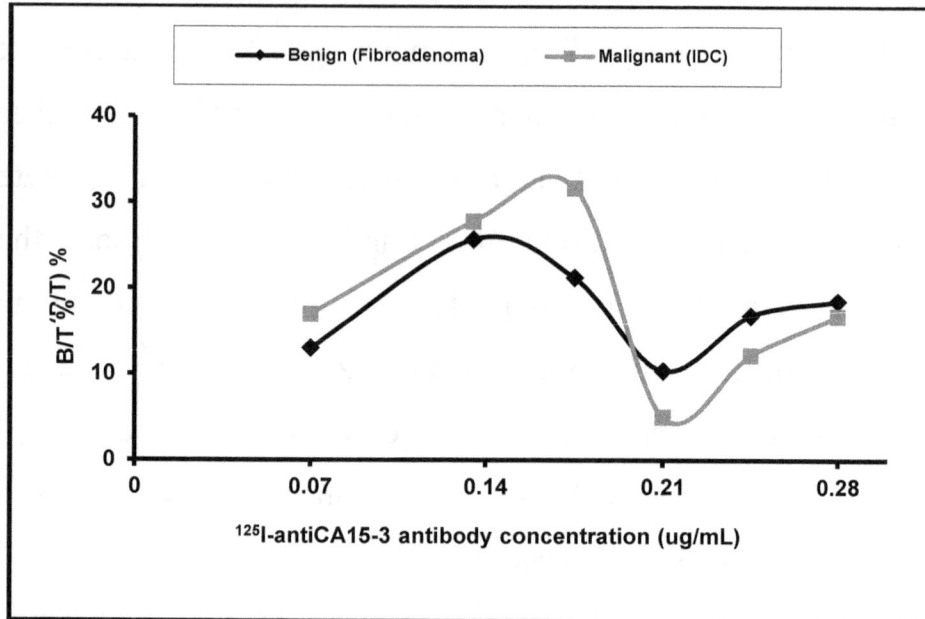

Figure (3.5): Effect of ^{125}I-anti CA15-3 antibody concentration on its binding with partially purified CA15-3 from breast tumors. (All other details are explained in the text).

Optimum pH

Figure (3-6) shows the effect of increasing pH on the binding of ^{125}I-anti CA15-3 antibody to its purified antigen. The results revealed that the optimum pH for (BI) and (MI) purified fractions for the binding with its antibody was 7.0. These results indicate that the binding was pH dependent.

The similarity in pH (7.0) suggests that the CA15-3 isolated from different sources of tissues either benign or malignant breast tissues homogenates possesses the same epitopes in both cases. That means the induction of protonation-deprotonation process [183] occurs within the same changed polar groups on the amino acid residues present in the binding domain. According to the results obtained, the pH of the buffer used in all subsequent experiments was adjusted to pH 7.0.

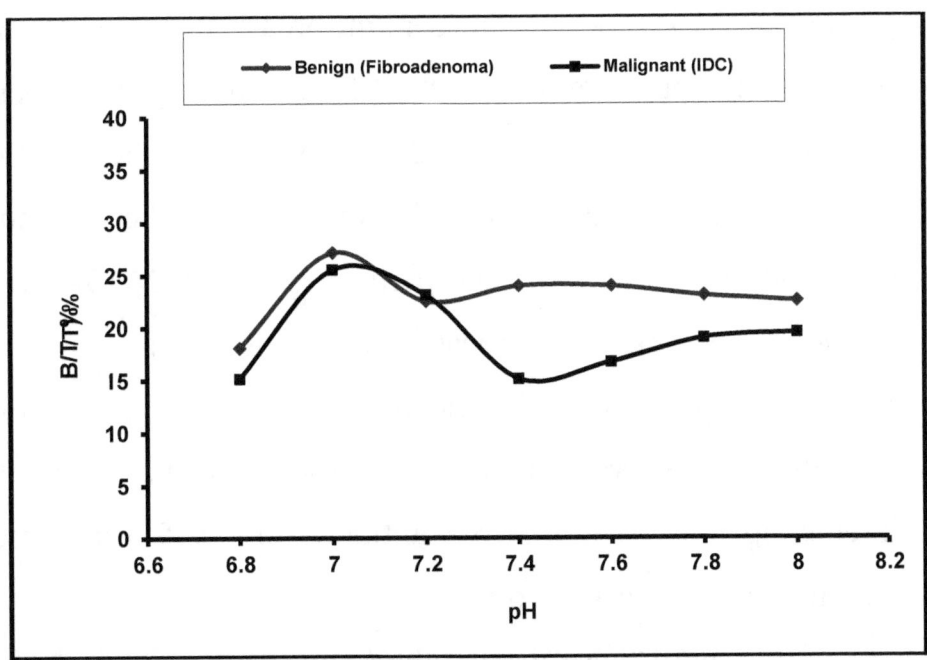

Figure (3.6): pH effect on the binding of ^{125}I-anti CA15-3 antibody with partially purified CA15-3 from breast tumors. (All other details are explained in the text)

Optimum Temperature

The temperature dependency of the isolated CA15-3 binding to its antibody ^{125}I-anti CA15-3 was investigated.

Figure (3.7) show the optimum temperatures on the binding of ^{125}I-anti CA15-3 antibody was 37°C with partially purified CA15-3 (BI) and 15°C with partially purified CA15-3 (MI).

The difference of the temperature between crude and purified CA15-3 occurs in benign breast tumors, i.e. the optimum temperature was 45°C of the binding of ^{125}I-anti CA15-3 antibody to crude CA15-3 while in the purified CA15-3 (BI) was 37°C. On the other hand, the optimum temperature in both crude and partially purified CA15-3 from premenopausal malignant breast tumors was 15°C.

The temperature dependency of the binding suggests that the whole process is controlled by diffusion of the interacting of ^{125}I-anti CA15-3 antibody to CA15-3 in benign and malignant breast tumors [184].

In view of these results, the temperatures (37 & 15 °C) for both benign and malignant breast tumors were used in all subsequent experiments.

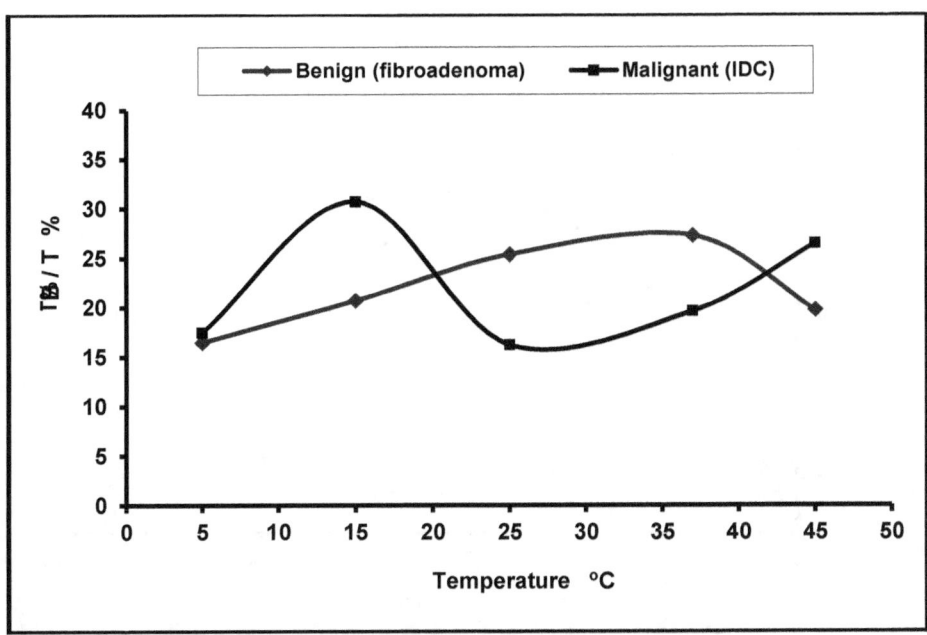

Figure (3.7): Effect of the temperature on the binding of ^{125}**I-anti CA15-3 antibody with partially purified CA15-3 from breast tumors. (All other are explained in the text)**

The Effect of Incubation Time

Figure (3.8) shows the time required for the highest binding of ^{125}I-anti CA15-3 antibody to partially purified CA15-3 in (BI) and (MI) was 90 min at 37 and 15 °C respectively.

In view of these results, the incubation time used in all subsequent experiments was 90 min.

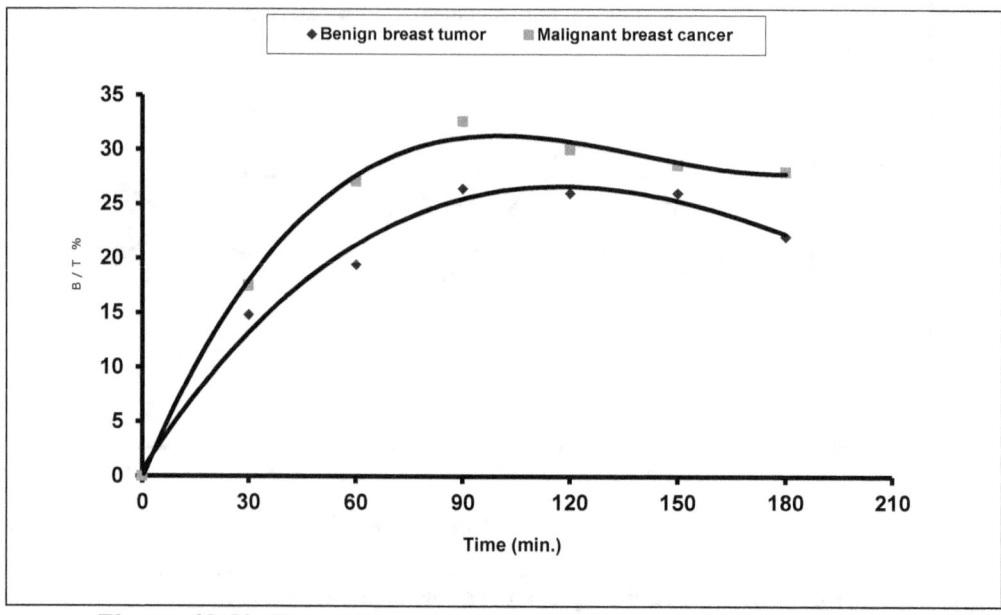

Figure (3.8): Time dependence of ^{125}I-anti CA15-3 antibody binding with partially purified CA15-3 from breast tumors. (All other details are explained in the text)

Stability of CA15-3 at −20 oC

The crude and isolated fractions of CA15-3 from malignant breast tumors were stored at −20 °C during the experiments. It was carried out in order to study the stability of CA15-3 and check their efficiencies of the binding through out the storage period. The results showed that CA15-3 of crude fraction was more stable than the isolated fractions as shows in figure (3-9). This result is in agreement with Al-Atrakchi observations [185].

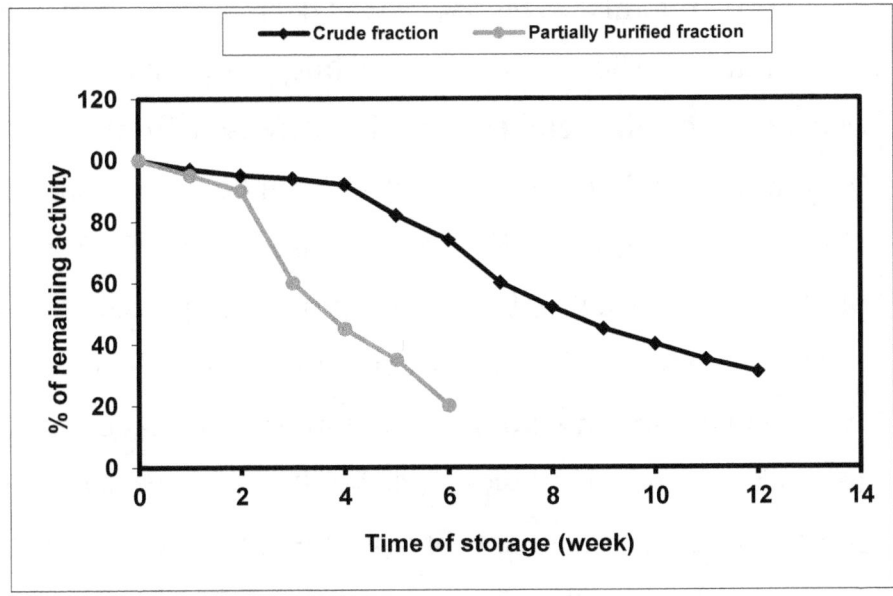

Figure (3.9): Stability of partially purified and crude CA15-3 upon storage at *Abstract*

Kinetic and thermodynamic parameter associated with the binding of ^{125}I-anti CA15-3 antibody to partially purified CA15-3 in both cases, benign and malignant breast tumors were investigated.

It was shown that the reaction in all studied cases follow pseudo-first order reaction kinetics. The maximum binding (B_{max}) of partially purified CA15-3 in benign breast tumors (Fibroadenoma) was 10.48×10^{-3} mg.mL^{-1} after 90 minutes incubation at 37°C, while the (B_{max}) of partially purified CA15-3 in malignant breast tumors (IDC) was 13.38×10^{-3} mg.mL^{-1}. The (B_{max}) was decreased with increasing temperature. The values of affinity constant (K_a) were dependent on the temperature, K_a increased from 14.18 mg^{-1}.mL at 5°C to 31.65 mg^{-1}.mL at 45°C in benign breast tumors (Fibroadenoma), while K_a was increased from 13.87 mg^{-1}.mL at 5°C to 23.81 mg^{-1}.mL at 45°C in premenopausal malignant breast tumors (IDC). The association constant K_{+1} increased with temperature in benign breast tumors (Fibroadenoma). On the other hand, K_{+1} was independent of temperatures in premenopausal malignant breast tumors (IDC). The Van't Hoff plot demonstrated a linear relationship between K_a and 1/T, using the partially purified CA15-3 in benign and malignant tumor homogenate. Arrhenius plot indicate that there was a linear-relationship between log K_{+1} and 1/T. The transition state thermodynamic parameters (Ea, ΔH^*, ΔG^*, ΔS^*) for the formation of (^{125}I-antiCA15-3 antibody /CA15-3) were determined.

Chapter Four

Kinetic and thermodynamics of Binding CA-15-3 to its antibody

Introduction

The specific reaction between an antibody (Ab) and an antigen (Ag) is usually driven by electrostatic forces between oppositely charged amino acids, hydrogen bonding, and hydrophobic interactions. The equilibrium reaction, termed "biospecific interaction", is characterized by the affinity of reactants to form Ag-Ab complex [186].

Kinetic studies supplement the information for differences between the initial, final states of each reactant and an intermediate activated complex, (i.e, the pathway taken by the reactants reach the final product) [187]. On the other hand, thermodynamics of the binding describes the system in its initial, final states. Using kinetic and equilibrium data also determined thermodynamic formation constant.

Al-Mudhuffar et.al, have many studies on the kinetic and thermodynamic of protein-protein interaction in human breast tissue, like kinetic and thermodynamic of purified steroid receptor of malignant breast tumors with hormone [188], also kinetic and thermodynamic studies on the binding of lectin in human malignant breast to glycoprotein [189].

In this chapter, the basic mathematical analysis was described and used to explain the mechanism through kinetics of binding of CA15-3 from both breast tumor homogenates (fibroadenoma and Infiltrating ductalcarcinoma) to its antibody to form (^{125}I-anti CA15-3 antibody / CA15-3) complex in partially purified fraction.

Materials and Methods

4.1. Materials

4.1.1. Chemicals

All chemical and reagents mentioned in section (2.1.1) in chapter two were used in the experiments of this chapter.

4.1.2. Instruments

All instruments that were described in section (2.1.2) in chapter two were used in the experiments of this chapter.

4.2. Methods

4.2.1. Kinetic Studies

4.2.1.1. The Time-Course of the Binding of 125I-anti CA15-3 Antibody with CA15-3 in Breast Tumor Homogenate

1. One hundred microliters of partially purified CA15-3 from benign breast tumor (fibroadenoma) and premenopausal malignant breast tumor (Infitrating ductal carcinoma, IDC) containing (150 and 100 $\mu g.mL^{-1}$ protein) respectively, were added to (20 and 25 μL) of ^{125}I-anti CA15-3 antibody containing (0.140 and 0.175 $mg.mL^{-1}$) respectively.

2. The volume of reaction were completed to 500 μL with PBS buffer pH 7.0.

3. All tubes were incubated at 37°C at different time intervals (30, 60, 90, 120, 150, 180) min.

4. Steps 3, 4, 5 and 6 of experiment (2-4-2-1) were repeated.

5. To determine the time-course of partially purified CA15-3 binding to ^{125}I-anti CA15-3 antibody at different temperatures, step 1,2,3 and 4 in the same experiment were repeated at different temperatures 5, 15, 25 and 45C°.

Calculation

The values of (B/T)% were calculated as described in section (2.4.1) and plotted against incubation time at each temperature for both types of homogenates.

4.2.1.2. Determination of Kinetic Parameters of 125I-Anti CA 15-3 Antibody Binding with Partially Purified CA 15-3 in Benign and Malignant Breast Tumors

Determination of the affinity constant (K_a) and the maximal binding capacity (B_{max}) of:

A. Partially Purified CA15-3 in Benign Breast Tumor Homogenate Binding with ^{125}I-Anti CA15-3 Antibody

1. One hundred microliters of partially purified CA15-3 from benign breast tumor (Fibroadenoma) containing (150 $\mu g.mL^{-1}$ protein) were added to increasing volumes (4, 8, 12, 16, 20 and 24 μL) of ^{125}I-anti CA15-3 antibody containing (0.0280, 0.0560, 0.0841, 0.1121, 0.1402 and 0.1684 $mg.mL^{-1}$) to each assay tube. The final volume of each assay tube was completed to 500 μL with PBS buffer pH 7.0.

2. All tubes were incubated for 90 min at 37°C.

3. Steps 3, 4, 5 and 6 in experiment (2.4.2.1) were repeated at different temperatures (5, 15, 25 and 45°C).

4. The time of incubation required to reach the equilibrium state are reported in table (4-1) according to the following:

Table (4.1): The time of incubation for benign and malignant breast tumor homogenate at different temperatures

Temp. °C	Time (min.)	
	Benign breast tumor homogenate (Fibroadenoma)	Malignant breast tumor homogenate (IDC)
5	180	180
15	60	90
25	90	150
37	90	90
45	180	90

Calculations

1- The B/F ratio was computed for each tube, where:

B: is the bound radioactivity (mean counts in c.p.m), which represent the formation of (^{125}I-anti CA15-3 /CA15-3) complex.

F: is the free radioactivity (mean counts in c.p.m.), which represents the (unbound or unreacted), ^{125}I-anti CA15-3 antibody.

T: is the total activity (mean counts in c.p.m.)

$$F = T \text{ (total counts) - B (bound radioactivity)}$$

2- The concentration of (^{125}I-anti CA15-3/CA15-3) complex in $mg.mL^{-1}$ which found after time (t) was calculated from the following equation:

$$B(mg.mL^{-1}) = \frac{B(c.p.m)}{T(c.p.m)} \times \text{Concentration of }^{125}I - \text{anti } CA15-3 \text{ antib}$$

the incubation medium in $mg.mL^{-1}$

3- The affinity constant and maximal binding capacity were determined according to Scatchard equation [190,191].

$$\frac{B}{F} = \frac{1}{K_d} \times (B_{max} - B)$$

$$K_a = \frac{1}{K_d} = \frac{K_{+1}}{K_{-1}}$$

Where: K_a = affinity constant

K_d = dissociation constant

B_{max} = maximal binding capacity

The value of affinity constant of the binding Ka at each temperature can be calculated from the slop of the straight line in figure (4.2), while the value of the total concentration of CA15-3 (B_{max}) in breast tumor homogenate for each group was calculated from the intercept of the x-axis.

B. Partially Purified CA15-3 in Human Malignant Breast Tumor Homogenate Binding with ^{125}I-Anti CA15-3 Antibody

1. One hundred microliters of partially purified CA15-3 from premenopausal malignant breast tumor (IDC) containing (100 $\mu g.mL^{-1}$ protein) were added to increasing volumes (5, 10, 15, 20, 25 and 30 μL) of ^{125}I-anti CA15-3 antibody containing (0.035, 0.070, 0.105, 0.140, 0.175 and 0.210 $mg.mL^{-1}$) to each assay tube. The final volume of each assay tube was completed to 500 μL with PBS buffer pH 7.0.

2. All tubes were incubated for 90 min at 15°C

3. Steps 3, 4, 5 and 6 in experiment (2.4.2.1) were repeated at different temperatures (5, 25, 37 and 45 °C).

4. The times of incubation required to reach the equilibrium state are reported in table (4.1).

Calculations

The method outlined in experiment (4.3.2.A) was followed exactly to obtain the values of K_a and B_{max} at each temperature as shown in figure (4.3).

4.3. The Thermodynamic Studies of 125I-Anti CA15-3 Antibody Binding to Partially Purified CA15-3 in Benign and Malignant Breast Tumors

The same steps mentioned in section (4.2.1.1) and (4.2.1.2) were performed using the dialyzable protein fraction of benign and malignant breast tumor homogenate from fibroadenoma and (IDC) as the partially purified CA15-3 source.

Calculation

1. The thermodynamic parameters of standard state were obtained from Van't Hoff plot, the values of the natural

logarithm of equilibrium constant (affinity constant K_a) obtained at different temperatures were plotted against the reciprocal values of the absolute temperature in Kelvin (1/T), according to the following equation:

$$\ln K_a = \frac{\Delta S^\circ}{R} - \frac{\Delta H^\circ}{RT}$$

Where:

ΔH° = the enthalpy change of the standard state.

ΔS° = the entropy change of the standard state.

R = the gas constant ($8.314\ J.K^{-1}.mol^{-1}$).

ΔH° value obtained from the slop of a linear relationship of the plot.

The change in Gibbs free energy of the standard state ΔG° was obtained from the following equation:

$$\Delta G^\circ = -RT\ Ln\ K_a$$

Where Ka is the affinity constant, while the standard state entropy change was obtained from [192]:

$$\Delta S^\circ = \frac{\Delta H^\circ - \Delta G^\circ}{T}$$

2. The thermodynamic parameters of the transition state were obtained from Arrhenius plot of Ln K_{+1} values against (1/T) values, that given a linear relationship according to the following equation:

$$\text{Ln } K_{+1} = \text{Ln } A - \left(\frac{E_a}{RT}\right)$$

Where:

A: Arrhenius constant .

The values of activation energy (E_a) of the binding reaction can be determined from the slop of the straight line.

The enthalpy of transition state ΔH^* was obtained from:

$$\Delta H^* = E_a - RT$$

Transition state of free energy change ΔG^* is calculated from the following equation:

$$\Delta G^* = -RT \text{ Ln} K_{+1} + RT \text{ Ln} \frac{KT}{h}$$

where K and h were Boltzmann and Plank's constant which equal

$(1.38 \times 10^{-23} \text{ J.K}^{-1})$, $(6.62 \times 10^{-34} \text{ J.sec}^{-1})$ respectively.

The change in entropy of the transition state AS* is calculated from the following equation:

$$\Delta S^* = \frac{\Delta H^* - \Delta G^*}{T}$$

Results and Discussion

Kinetic Studies

The Time-Course of the Binding of 125I-anti CA15-3 Antibody with CA15-3 in Breast Tumor Homogenate

Figure (4.1.A & B) shows the time – course of the formation of (^{125}I-anti CA15-3 /CA15-3) complex at five different temperatures (5, 15, 25, 37 and 45°C) of partially purified CA15-3 from benign and malignant breast tumors homogenates samples.

The concentration of (^{125}I-anti CA15-3/CA15-3) complex formed after time (t) was calculated from the following equation:

$$[^{125}\text{I-antiCA15-3/CA15-3}] \text{ in mg.mL}^{-1} \text{ after time (t)} = \frac{\text{Count (c.p.m.) of } ^{125}\text{I-antiCA15-3 specifically bound after time (t)}}{\text{Total counts (c.p.m.) of } ^{125}\text{I-anti CA15-3 used in the incubation}} \times \text{Concentration of } ^{125}\text{I-antiCA15-3 in the incubation (mg.mL}^{-1})$$

The results of time-course pattern at different temperatures indicated that the equilibrium binding studies is temperature and time dependent process. In case premenopausal malignant breast tumor (IDC) the maximum binding occurs at 15°C (after incubation for 90 minutes), while in benign breast tumors (fibroadenoma) the maximum binding occurs at 37°C at the same incubation time. This is may be due to the different source of CA15-3. Several authors studied the time – course of purified steroid receptors of malignant breast tumors[188], others studied the time – course on the binding of lectin in human malignant breast to glycoprotein [189], these studies revealed that the time-course must be done to find the maximum binding at different incubation time as a step to prepare the kinetic and thermodynamic studies.

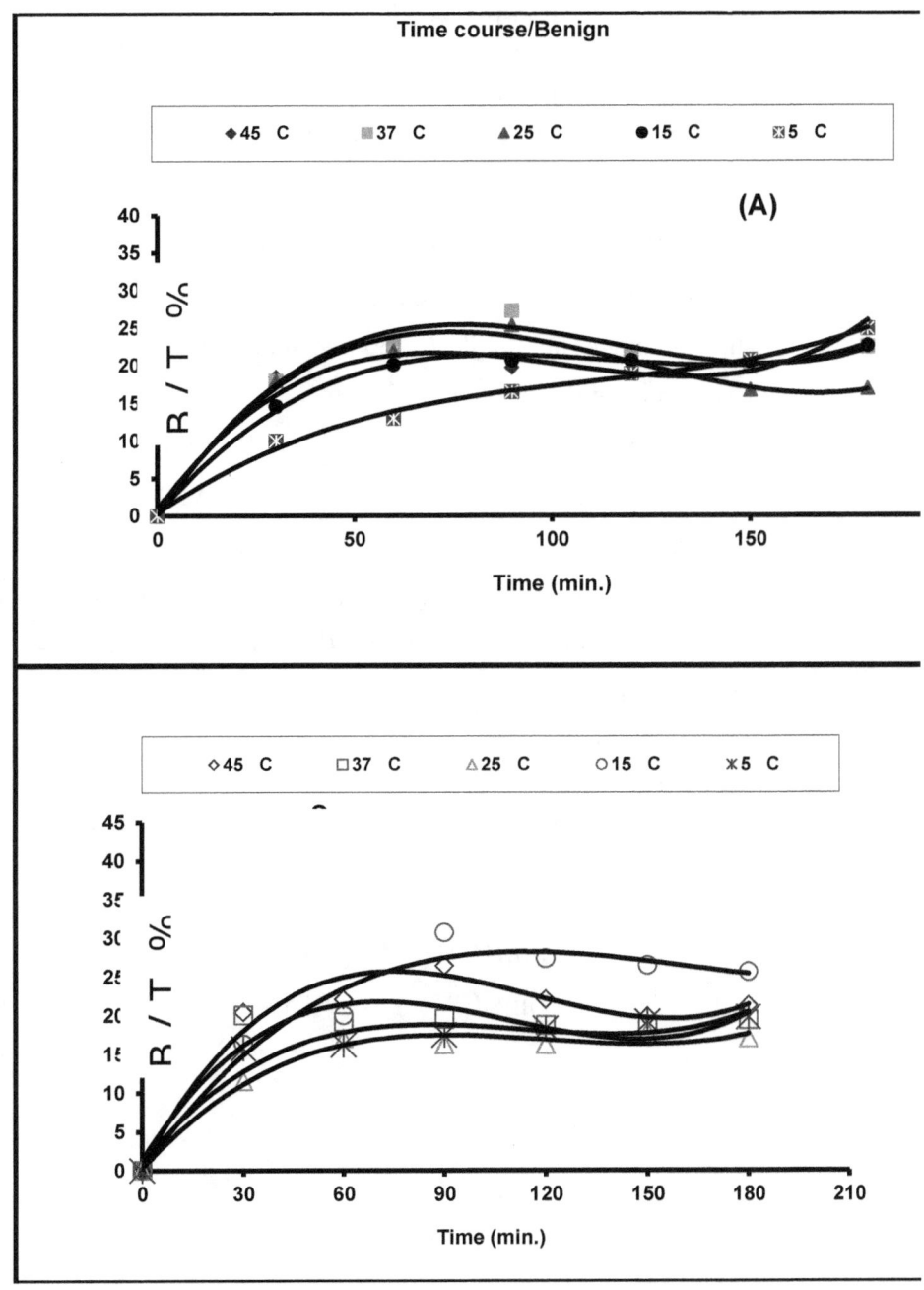

Time course/Benign

| ◆45 C | ■37 C | ▲25 C | ●15 C | ☒5 C |

(A)

| ◇45 C | □37 C | △25 C | ○15 C | ✳5 C |

Figure (4.1): Time-Course of ^{125}I-anti CA15-3 binding to partially purified CA15-3 in:

(A)   Benign tumor (Fibroadenoma) tissue homogenate.
(B)   Malignant tumor (IDC) tissue homogenate.
(All other details are explained in the text).

Determination of Kinetic Parameters of 125I-Anti CA15-3 Antibody Binding with Partially Purified CA15-3 from Benign and Malignant Breast Tumors

The time course of (^{125}I-anti CA15-3/CA15-3) complex formation was carried out to describe the kinetic parameters of the binding. The simplest proposed model representing this interaction is:

$$^{125}\text{I-antiCA15-3} + \text{CA15-3} \underset{K_{-1}}{\overset{K_{+1}}{\rightleftharpoons}} [^{125}\text{I-antiCA15-3/CA15-3}]$$

Where:

K_{+1}: is the association rate of ^{125}I-anti CA15-3 to /or CA15-3.

K_{-1}: is the dissociation rate of (^{125}I-anti CA15-3/CA15-3) complex formed.

At equilibrium:

$$K_a = \frac{[^{125}\text{I} - \text{antiCA15} - 3/\text{CA15} - 3]}{[^{125}\text{I} - \text{antiCA15} - 3][\text{CA15} - 3]} \quad \ldots\ldots\ldots\ldots(2)$$

$$K_d = \frac{[^{125}\text{I} - \text{antiCAl5} - 3][\text{CA15} - 3]}{[^{125}\text{I} - \text{antiCAl5} - 3/\text{CA15} - 3]} \quad \ldots\ldots\ldots\ldots(3)$$

Thus:

$$K_a = \frac{1}{K_d} = \frac{K_{+1}}{K_{-1}} \ldots\ldots\ldots\ldots\ldots\ldots\ldots\ldots\ldots\ldots (4)$$

Where:

The value K_a and maximal binding capacity (B_{max}). Were calculated from Scatchard plot at five different temperatures at incubation time of 90 minutes, figure (4-2) and (4-3).

It is clear from table (4-2), that the affinity constant (K_a) is depended on the type of the tumor (i.e., benign or malignant) and on the temperature. K_a increased with increased temperature for the same tumor (Fibroadenoma), K_a increased from 14.18 $mg^{-1}.mL$ at 5°C to 31.65 $mg^{-1}.mL$ at 45°C. Whereas the values of dissociation constant (K_d) was calculated by using equation (4), which show that the lowest K_d value of (^{125}I-anti CA15-3/CA15-3) complex occurs at 45°C at time of incubation 180 minutes.

The concentration of CA15-3 in partially purified fractions of (Fibroadenoma) was determined to be 10.48×10^{-3} $mg.mL^{-1}$ and the maximum binding (B_{max}) occurred after 90 minutes incubation at 37 °C. While in the same table the maximum K_a value for the binding ^{125}I-anti CA15-3 antibody with CA15-3 present in partially purified fraction of (IDC) occurred at 15°C and it was increased with temperature in the following order: 5 >15 >25 >37 > 45 °C.

The lowest K_d value of (^{125}I-anti-CA15-3 /CA15-3) complex occurs at 45 °C at the time of incubation.

Scatchard plot analysis gave straight line as shown in figure (4.2) and (4.3) indicating that the (^{125}I-anti CA15-3/CA15-3) complex is directed against the same epitopes on CA15-3 molecules. On the other hand, the maximum binding occurred at 15°C and was 13.38×10^{-3} mg.mL^{-1} also shows that the (B_{max}) decreased with increasing temperatures of incubation.

Table (4-2): The Kinetic parameter of ^{125}I-anti CA15-3 antibody binding to partially purified CA15-3 in breast tumor homogenate. (All other details are explained in the text).

Temp °C	Benign breast tumors (Fibroadenoma)			Malignant breast tumors (IDC)		
	Binding Capacity $B_{max} \times 10^{-3}$ (mg.mL^{-1})	K_a (mg^{-1}.mL)	$K_d \times 10^{-2}$ (mg.mL^{-1})	Binding Capacity $B_{max} \times 10^{-3}$ (mg.mL^{-1})	K_a (mg^{-1}.mL)	$K_d \times 10^{-2}$ (mg.mL^{-1})
5	9.22	14.18	7.05	10.82	13.87	7.21
15	8.05	16.73	5.98	13.38	20.78	4.81
25	9.02	16.38	6.10	12.57	20.84	4.79
37	10.48	18.66	5.36	9.63	22.22	4.50
45	6.67	31.65	3.16	11.67	23.81	4.20

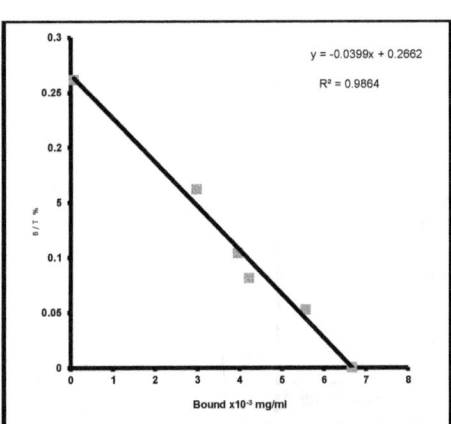

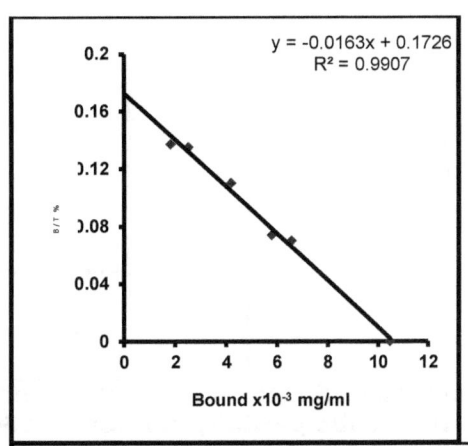

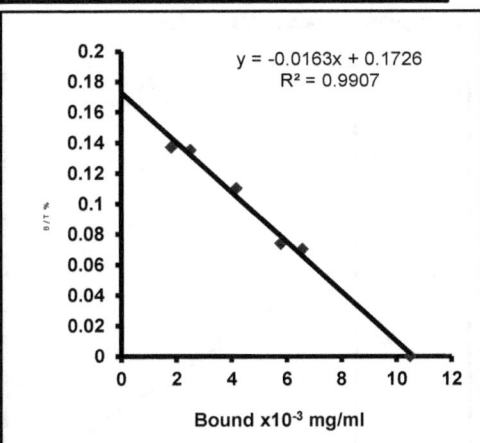

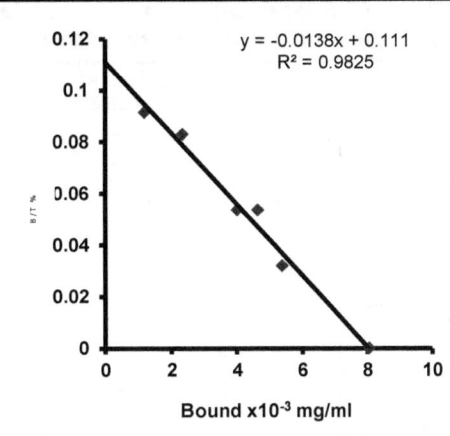

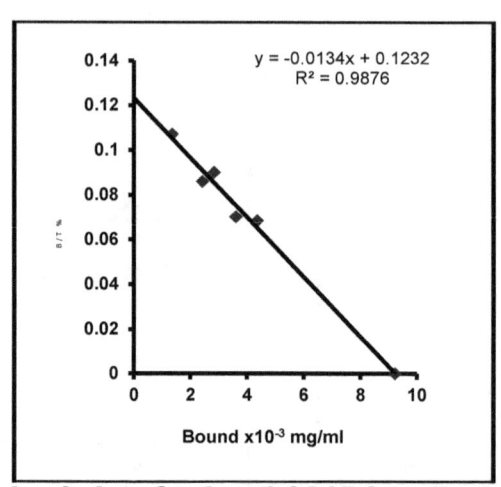

Figure (4-2): Scatchard plot of ^{125}I-anti CA15-3 antibody binding to the partially purified CA15-3 in benign breast tumors (Fibroadenoma) at five different temperatures. All details are explained in the text.

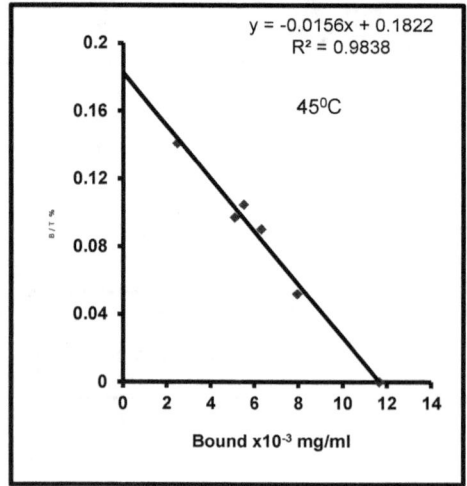

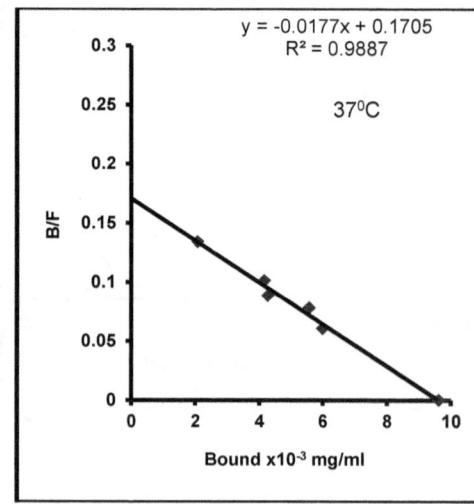

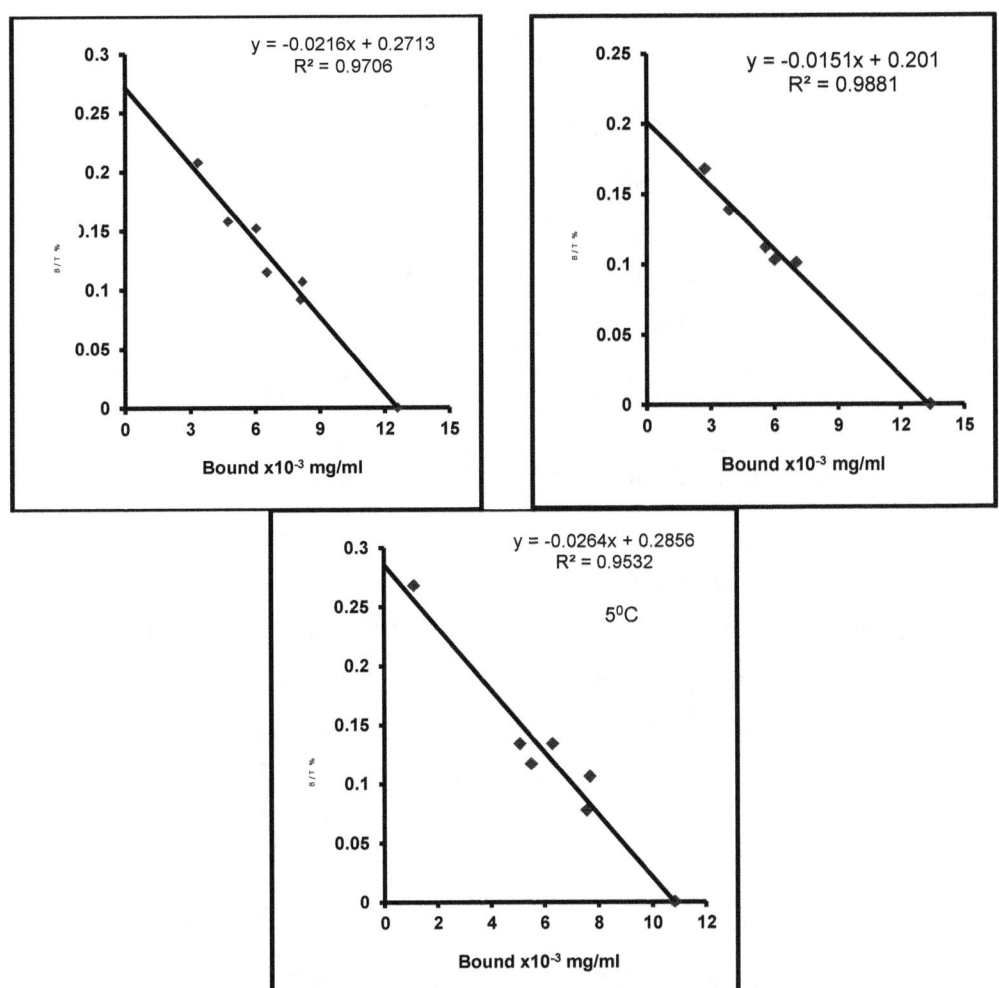

Figure (4.3): Scatchard plot of ^{125}I-anti CA15-3 antibody binding to the partially purified CA15-3 in Malignant breast tumors (IDC) at five different temperatures. All details are explained in the text.

However, the time-course data shown in figure (4-1) could be used to determine the reaction order of CA15-3 binding to its specifically ^{125}I-anti CA15-3 using the following equation [193]:

$$Ln[AbAg]_e \left[\frac{[Ab]_t - [AbAg]_t [AbAg]_e / [Ag]_t]}{[Ab]_t [AbAg]_e - [AbAg]_e} \right] = K_{+1}t \left[\frac{[Ab] + [Ag]_t - [AbAg]_e}{[AbAg]_e} \right] \quad ......(5)$$

Where:

k_{+1} : is the kinetic association constant in $mg^{-1}. min^{-1}. mL$.

$[AbAg]_e$: is the concentration of (^{125}I-antiCA15-3/CA15-3)complex   formed at equilibrium.

$[AbAg]_t$: is the concentration of (^{125}I-antiCA15-3/CA15-3) complex after time (t).

$[Ab]_t$: is the total concentration of ^{125}I-anti CA15-3 antibody in $mg. mL^{-1}$.

$[Ag]_t$: is the total concentration of CA15-3 in $mg. mL^{-1}$.

Equation (5) represents the second order kinetics, but the percent of binding was in some cases, small and most labeled antibody remains free and only small fraction binds even at equilibrium, i.e , $[Ab]_t \gg [AbAg]_e$

Thus :

$$[Ab]_t \gg \frac{[AbAg]_t[AbAg]_e}{[Ag]_t}$$

So that the following equation [187] could be used in order to fit the pseudo-first order kinetics:

$$Ln\frac{[AbAg]_e}{[AbAg]_e - [AbAg]_t} = K_{+1}t\frac{[Ab]_t[Ag]_t}{[AbAg]_e} \quad ....6)$$

On the other hand, figure (4-4) and (4-5) show the plot of $\ln\dfrac{[AbAg]_e}{[AbAg]_e - [AbAg]_t}$ Against time (t) in both benign and malignant breast tumors, which give a straight line with a slope equal to the observed value of first rate constant K_{bos} in min^{-1}. The rate constant (k_{+1}) in mg^{-1}. mL. min was calculated at five different temperatures by using the following equation [194]

$$K_{obs} = K_{+1}\frac{[^{125}I - antiCA15 - 3]_t[CA15 - 3]_t}{[^{125}I - antiCA15 - 3/CA15 - 3]_e} \quad ....(7)$$

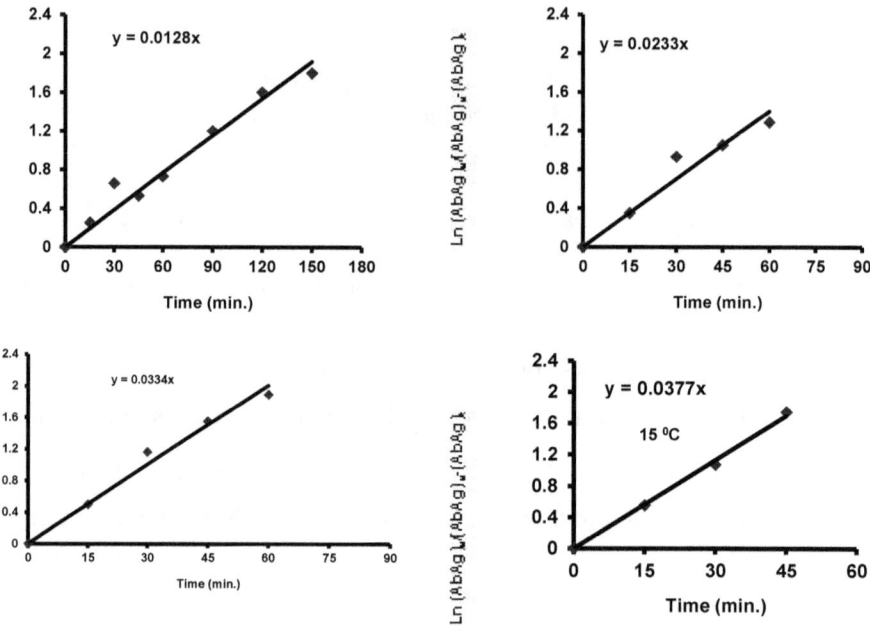

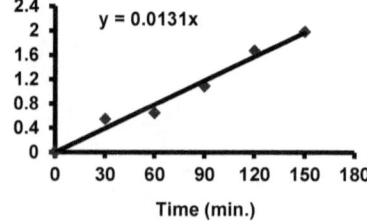

Figure (4.4): Kinetics of ^{125}I-anti CA15-3 antibody binding to partially purified CA15-3 in benign breast tumors (Fibroadenoma). All details are explained in the text.

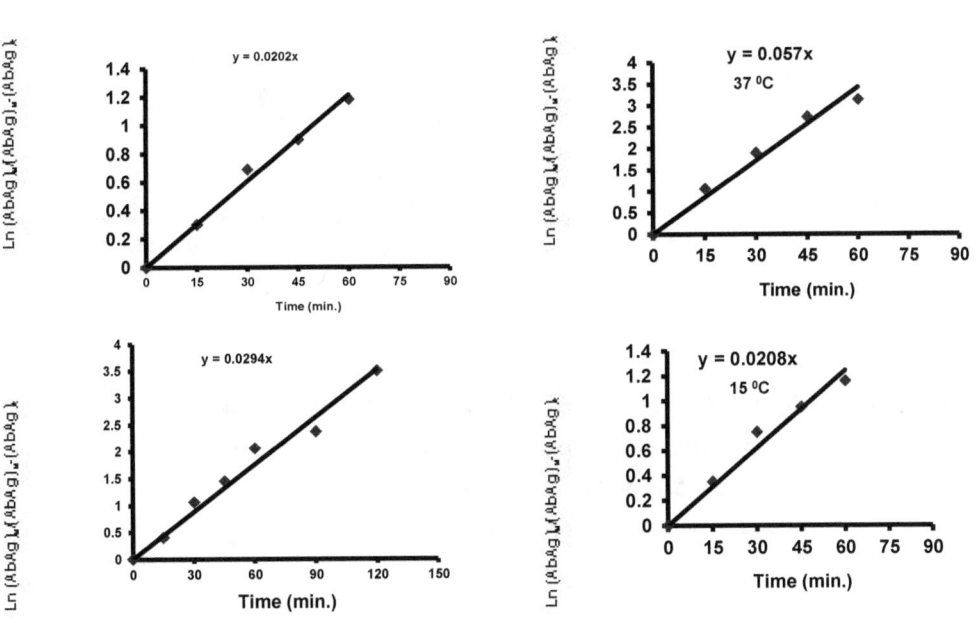

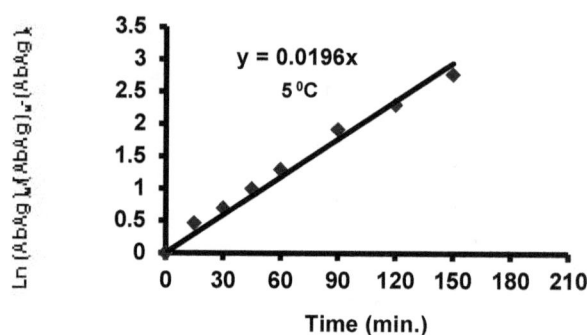

$$y = 0.0196x$$

5 °C

Ln [(AbAg)∞/(AbAg)∞-(AbAg)ₜ]

Time (min.)

Figure (4.5): Kinetics of ¹²⁵I-anti CA15-3 binding to partially purified CA15-3 in malignant breast tumors (IDC). (All details are explained in the text).

The value of k_{-1} at five temperatures was calculated by using equation (4). Whereas, the half-life time of association $(t \, \frac{1}{2})_{ass.}$, Which represented the time needed for the formation of half amount of the complex at equilibrium was determined from the concentration of the complex at equilibrium and the time-course curve. The half-life time of dissociation $(t \, \frac{1}{2})_{diss.}$, was calculated from the following relation:

$$(t_{1/2})_{diss.} = \frac{\ln 2}{k_{-1}} = \frac{0.693}{k_{-1}}$$

$$(t_{1/2})_{ass.} = \frac{\ln 2}{k_{obs}} = \frac{0.693}{k_{+1}}$$

The value of k_{obs} , k_{+1}, k_{-1}, $(t \, \frac{1}{2})_{ass.}$, $(t \, \frac{1}{2})_{diss.}$ at five different temperatures are summarized in table (4.3). Data analysis in this table shows that highest rate for the association reaction k_{-1}, in benign breast tumors (Fibroadenoma) and malignant breast tumors (IDC) occurs at 37°C and 15°C respectively, while the lowest rate occurs at 45°C. This means the dependence of reaction rate on temperature (Table 4.3) that also shows the values of the rate constant for the reverse reaction k_{-1} calculated from equation (4). Results show that the rate of dissociation of 125I-anti CA15-3 antibody, from its CA15-3 is temperature independent.

Table (4.3): The effect of temperature on the kinetic parameters of ^{125}I-anti CA15-3 binding to partially purified CA15-3 in benign and malignant breast tumors at five different temperature.

Temp. °C	$k_{obs} \times 10^{-3}$ (min^{-1})		K_{+1} mg^{-1}.ml.min^{-1}		$k_{-1} \times 10^{-1}$ (min^{-1})		$(t_{1/2})_{ass}$ (min)		$(t_{1/2})_{diss}$ (min)	
	Benign (Fibroadenom)	Malignant (IDC)	Benign (Fibroadenoma)	Malignant (IDC)	Benign (Fibroadenoma)	Malignant (IDC)	Benign (Fibroadenoma)	Malignant (IDC)	Benign (Fibroadenoma)	Malignant (IDC)
5	12.8	20.20	48.69	45.81	15.38	34.68	54	34	45	20
15	23.3	57.00	60.65	116.16	32.50	73.75	30	12	21	9
25	33.4	24.9	93.98	35.48	57.37	18.07	21	28	12	38
37	37.7	20.30	103.54	46.61	61.89	22.43	18	34	11	31
45	13.1	19.60	35.40	21.34	24.96	10.58	53	35	28	66

Thermodynamic Parameters of Standard State

Figure (4.6) and (4.7) show Van't Hoff plot of the binding of ^{125}I-anti CA15-3 antibody to the partially purified CA15-3 in benign breast tumors (Fibroadenoma) and malignant breast tumors (IDC) respectively, at different temperatures (5 , 15 , 25 , 37 and 45 °C).

These figures revealed that the equilibrium binding constant (affinity constant) for CA15-3 to its antibody is a temperature dependent. The results indicated that $\Delta H°$, in general, had small values and their positive sign ascertains that the reaction was nearly endothermic. The $\Delta H°$ value in the case of the binding of ^{125}I-anti CA15-3 antibody to partially purified CA15-3 in benign breast tumors 12.71 KJ.mol^{-1} was higher than that in case of binding in malignant breast tumors (IDC) 6.7 KJ.mol^{-1}, so more energy is needed in case of benign breast tumor for the reaction (binding) to occur. The small positive value of $\Delta H°$ may indicate a favorable interaction between ^{125}I-anti CA15-3 antibody with partially purified CA15-3 in both cases.

These include the non-covalent interaction, which are fundamentally electrostatic in nature such as charge-charge, charge-dipole, dipole-dipole, charge-induced dipole, dipole-induced dipole interactions, and hydrogen bonds. The sum of these types of interactions can yield some stabilization to the folded structure of the complex [195].

The other values of thermodynamic parameters of standard state at five temperatures, such as $\Delta G°$ values and $\Delta S°$ values are summarized in table (4.4) and (4.5).

Table (4.4): Thermodynamic parameters at standard state of ^{125}I-anti CA15-3 to the partially CA15-3 in benign breast tumors (Fibroadenoma). (All other details are explained in the text).

Temp. °C	$\Delta H°$ KJ .moL^{-1}	$\Delta G°$ KJ .moL^{-1}	$\Delta S°$ J .mol^{-1}.K^{-1}
5	12.71	-36.87	137.20
15	12.71	-38.59	138.42
25	12.71	-39.88	138.10
37	12.71	-41.82	139.01
45	12.71	-44.30	143.30

Table (4.5): Thermodynamic parameters at standard state of ^{125}I-anti CA15-3 to the partially purified CA15-3 in malignant breast tumors (IDC). (All other details are explained in the text).

Temp. °C	$\Delta H°$ KJ .moL^{-1}	$\Delta G°$ KJ .moL^{-1}	$\Delta S°$ J .mol^{-1}.K^{-1}
5	6.70	-36.82	156.55
15	6.70	-39.11	159.06
25	6.70	-40.48	158.32
37	6.70	-42.27	157.97
45	6.70	-43.54	157.99

The negative values of $\Delta G°$ reflects the stability of the complex hence. The high affinity of the reactants. The high negative values of $\Delta G°$ for the binding reaction are controlled by high positive $\Delta S°$ values of the complex formed. So, our system is characterized by the sole contribution of $\Delta S°$ to the stability of the complex formed, which $\Delta H°$ has little or no effect [196]. Whereas, the negative values of $\Delta G°$ indicates that the reaction is spontaneous at the standard condition. On the other hand, the high positive of $\Delta S°$ suggest that the binding was entropically driven. Entropy has a driven force for the occurrence of the binding reaction, this indicates that the hydrophobic interactions played an important role in the stability of complex formation [197].

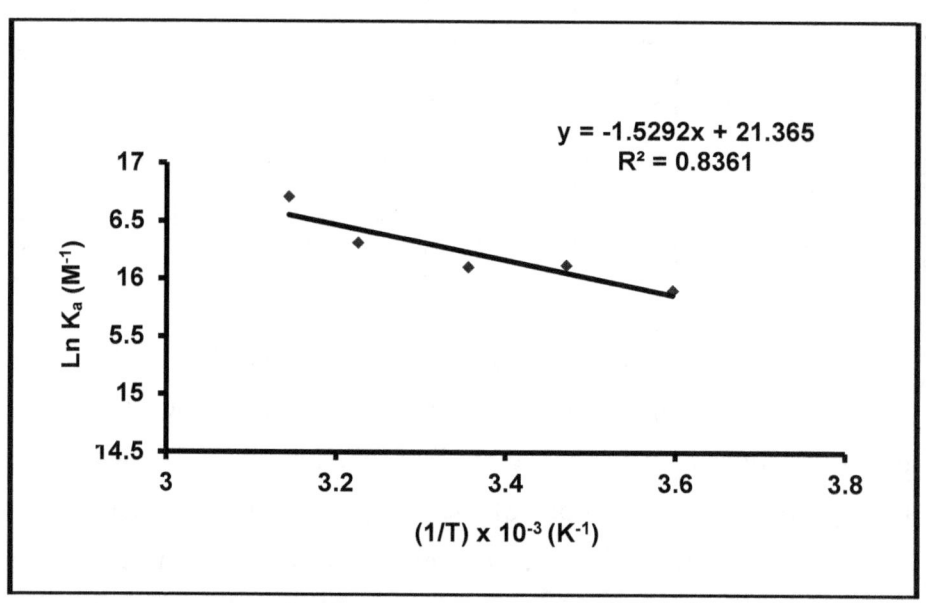

$$y = -1.5292x + 21.365$$
$$R^2 = 0.8361$$

Figure (4.6): Van't Hoff plot for the binding of ^{125}I-anti CA15-3 antibody to the partially purified CA15-3 in benign breast tumors (Fibroadenoma). All details are explained in the text.

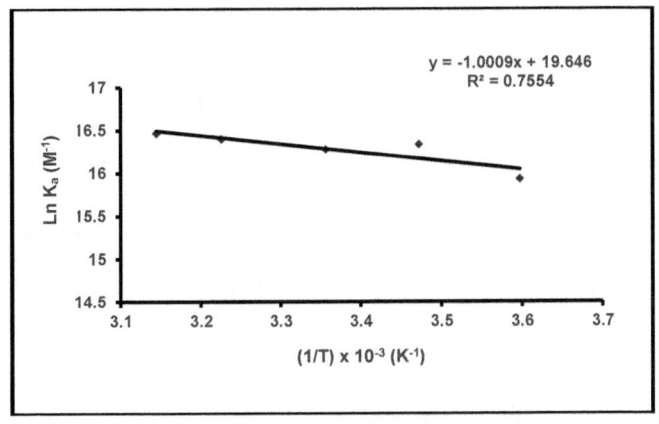

Figure (4.7): Van't Hoff plot for the binding of ^{125}I-anti CA15-3 antibody to the partially purified CA15-3 in malignant breast tumors (IDC). All details are explained in the text.

B. Thermodynamic Parameters of Transition State

Transition state theory postulated that the interaction of two substances to form the final product proceeds through the formation of an activated complex (transition state).

Consequently, the association of ^{125}I-anti CA15-3 antibody with its CA15-3 can be represented as follows:

$$^{125}I - antiCA15 - 3 + CA15 - 3 \rightarrow \left[^{125}I - antiCA15 - 3/CA15 - 3\right] \rightarrow \left[^{125}I - antiCA15 - 3/CA15 - 3\right]$$

$$\text{State}(A) \qquad\qquad \text{An Activated Complex} \qquad\qquad \text{Final Product}$$
$$\text{Transition State} \qquad\qquad\qquad \text{State}(B)$$

Thermodynamic parameters (ΔH^*, ΔG^* and ΔS^*) of the transition state were determined from the application of Arrhenius equation to the kinetic data. Figure (4.8) and (4.9) show Arrhenius plots for the binding of CA15-3 to its antibody, the slope of the line represents the activation energy (E_a) of the binding reaction, the linear relationship indicates the dependency of the association rate constant of the binding of CA15-3 to its antibody for benign and malignant breast tumors homogenate on temperature.

Table (4.6) and (4.7) show the values of thermodynamic parameters of the transition state (E_a, ΔH^*, ΔG^* and ΔS^*).

The high values of activation energy 9.96 KJ.mol^{-1} and 41.76 KJ.mol^{-1} of CA15-3 partially purified from benign and malignant breast tumors respectively, represents the required energy to overcome the energy barrier of the transition state for the formation of (^{125}I-anti CA15-3 antibody / CA15-3) complex. Also the value of activation energy is in accordance with the high positive values of ΔG^*, which indicates that the formation of the activated complex is a non-spontaneous process and requires a lot of energy (equal to E_a) to overcome the transition state energy barrier and giving the final product, whereas the high negative ΔS^* revealed that the activated complex had a more order structure than the reactants.

From the result obtained of the thermodynamic parameters in the transition state, it can be concluded that the positive values of ΔH^* and high positive values of ΔG^* are favorable to overcome the energy barrier of the transition state, the high negative values of ΔG^* is mainly attributed to the decrease in entropy of the transition state ($\Delta S^* < 0$).

In addition the positive values of ΔH^* show that the heat content of the activated complex is more than that in isolated species [193,198].

It is proposed that the formation of a complex occurs in the two steps. The first is the stabilization of the complex by hydrophobic interactions and second is the stabilization by short range interactions , such as electrostatic interaction, hydrogen bonding and Van der Waals interactions [199].

Hydrophobic interactions contribute to the complex stability via high positive entropy change ($\Delta S^* > 0$), while electrostatic interactions, hydrogen bonding and Van der Waals interactions contribute to the stability of the complex via negative entropy change ($\Delta S^* > 0$) [199,200].

The thermodynamic data indicate that the binding of ^{125}I-anti CA15-3 antibody to partially purified CA15-3 are entropy driven and in agreement with the concept that hydrophobic interaction play an important rote in the formation of (^{125}I-anti CA15-3 antibody / CA15-3) complex.

Table (4.6): Thermodynamic parameters at transition state of
^{125}I-anti CA15-3 antibody to the partially purified CA15-3 in benign breast tumors (Fibroadenoma). (All other details are explained in the text).

Temp. °C	Ea KJ . mol^{-1}	ΔH^* KJ . mol^{-1}	ΔG^* KJ . mol^{-1}	ΔS^* J .mol^{-1}. K^{-1}
5	9.96	7.65	58.94	-184.50
15	9.96	7.57	60.62	-184.20
25	9.96	7.48	61.72	-182.01
37	9.96	7.38	64.06	-182.84
45	9.96	7.32	68.62	-192.77

Table (4.7): Thermodynamic parameters at transition state of
^{125}I-anti CA15-3 antibody to the partially purified CA15-3 in malignant breast tumors (IDC). (All other details are explained in the text).

Temp. °C	Ea KJ . mol^{-1}	ΔH^* KJ . mol^{-1}	ΔG^* KJ . mol^{-1}	ΔS^* J .mol^{-1}. K^{-1}
5	41.76	39.45	59.08	-70.61
15	41.76	39.37	59.09	-68.47
25	41.76	39.28	64.14	-83.42
37	41.76	39.18	66.12	-86.90
45	41.76	39.12	70.00	-97.11

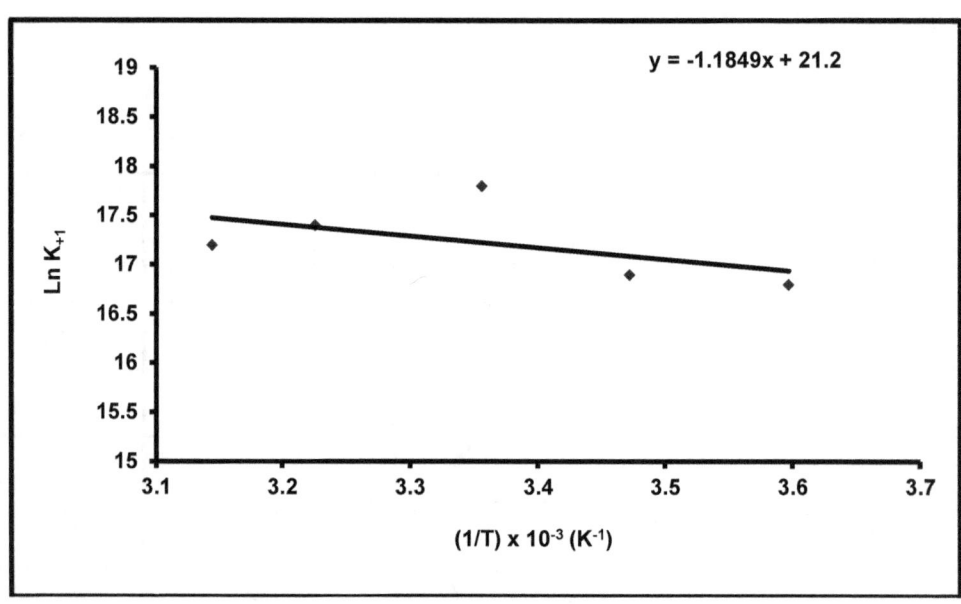

Figure (4-8): Arrhenius plot for the binding of [125]**I-anti CA15-3 to the partially purified CA15-3 in benign breast tumor (Fibroadnoma). All details are explained in the text.**

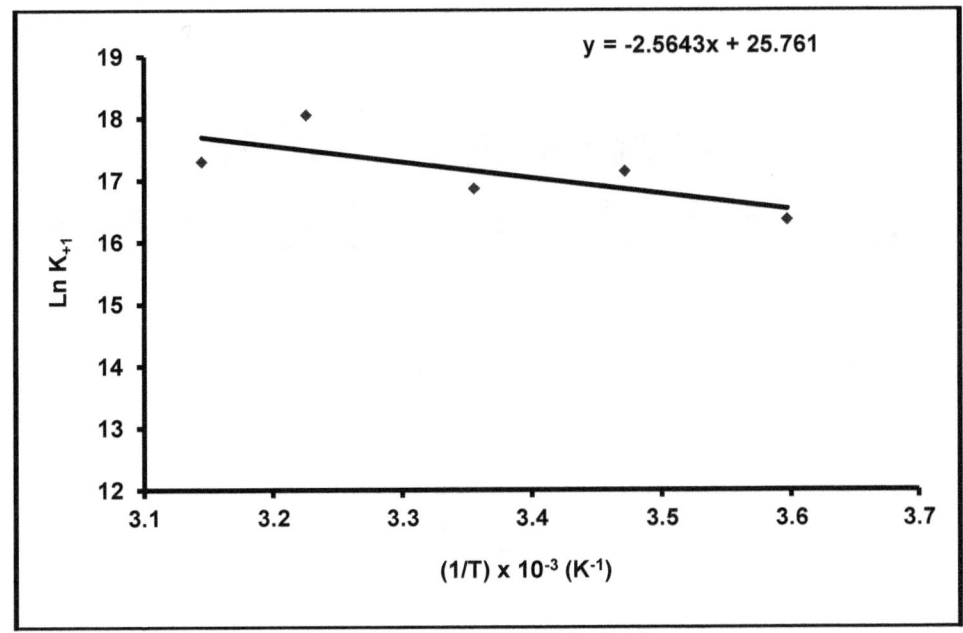

Figure (4.9): Arrhenius plot for the binding of ^{125}I-anti CA15-3 to the partially purified CA15-3 in malignant breast tumor (IDC). All details are explained in the text. Gel filtration technique was used to separate ^{125}I-anti CA 15-3 antibody bound to partially purified CA 15-3 using benign (Fibroadenoma) and malignant (IDC) breast tissue homogenate (as CA 15-3 source) from unbound (Free) ^{125}I-anti CA 15-3 antibody.

Chapter Five

characterization
of complexes of CA 15-3

The characterization of the complexes (^{125}I-anti CA 15-3 antibody/ CA15-3) from both benign and malignant breast tumors was carried out through the ultraviolet spectroscopic studies. Factors affecting the absorption properties of the two types of complexes such as pH, solvent polarity (solvent perturbation technique), spectrophotometric pH titration, and thermal stability in the presence of different concentrations of sodium chloride have been studied. pH titration of the two types of the complexes show that about (41.43%) and (44.29%) of histydyl residues are located on the surface of the two types of protein complexes (benign and malignant) respectively, while (40%) and (50%) of tyrosyl residues are buried interiorly in the complexes of (benign and malignant) respectively.

Molecules absorb light; the efficiency of absorption depend on both the structure and environment of the molecule making absorption spectroscopy a useful tool for characterizing both small and large molecule.

The ultraviolet absorption spectra of protein solutions in the region 250 to 310 nm are contributed from phenylalanyl, tyrosyl and tryptophanyl residues. But at the shorter wavelengths the contributions come from other groups such as histidyl residues and the peptide bond [169]. Changes in the environment of these chromophores can lead to alteration in the absorption spectrum, and the conformational changes of a protein may also involve environmental changes of its chromophoric groups [1201]. A variety of environmental changes (e.g. pH, temperature) can affect the absorption spectrum if the interaction of chromophore and perturbing agent affects the ground and excited states, the altered spectrum of the chromophore can be shifted to longer (red shift) or shorter (blue shift) wavelengths. The shift may or may not be accompanied by a change in intensity of the spectrum [170,202]. Saif-Alla, P.H., studied the UV spectra of h-PRL-antibody complex and CA15-3 molecule [203].

Interaction of h-CA 15-3 partially purified from benign (fibroadenoma) and malignant (IDC) tissues homogenate with its antibody is an example of protein-protein association. Although several new immunochemical techniques were developed to study such interactions [204,205], UV spectral remain as one of the most important methods in immunology because it provides a sensitive and quantitative measurements for the study of antibody structure and its specific ligand binding [206,208].

Very limited work concerning the physical properties of CA 15-3 specially those related to UV spectroscopy has been done, also the UV studies on CA 15-3 antibody interaction are not wide spread. Hence, this work is planned to study the association of the partially purified h-CA 15-3 and its antibody at different conditions.

Materials and Methods

5.1. Materials

5.1.1. Chemicals

All chemicals and reagents used in the experiments of this chapter were mentioned in section (2.1.1).

5.1.2. Instruments

The instruments used in this chapter are Shimadzu double beam UV- Visible spectrophotometer type 160, and instruments listed in section (2.1.2).

5.1.3 Buffers and Reagents

Buffers and reagents mentioned in section (2.1.6) are used in this chapter. Other additional solutions are indicators in each experiment.

5.2. Methods

5.2.1. Gel Filtration Technique for Separation of Free and Bound 125I -Anti CA 15-3 Antibody

5.2.1.1. Preparation of the Column

The dimensions of the column were (1x30 cm) chosen according to the equation in section (3.2.1.1).

5.2.1.2. Preparation of the Gel and Determination of Void Volume

The sepharose CL-4B was used to separate free and bound ^{125}I -anti CA 15-3 antibody, and was prepared as mentioned in section (3.2.1.3) and (3.2.1.4), the void volume was determined and found to be 10 mL.

5.2.1.3 Separation Procedure of (125I-Anti CA 15-3 Antibody/CA15-3) Complex

A) Partially Purified CA15-3 from Benign Breast Tumor (Fibroadenoma) and its Antibody ^{125}I - Anti CA 15-3

Reagents

Buffer PBS 0.15M, pH 7.0 containing 0.02% sodium azide was prepared as described previously in section (2.1.1.3).

Procedure

1- Partially purified CA 15-3 (475μL) containing (0.665 mg. mL^{-1}) was incubated with 120 μL of ^{125}I-anti CA 15-3 antibody (0.8412mg. mL^{-1}) and complete the reaction to a final volume of 700 μL with PBS buffer 0.15 M pH 7.0. The tubes were incubated for 90 min. at 37°C.

2- At the end of incubation, the mixture was applied to the surface of a sepharose CL-4B (1x30 cm) with a bed volume (23.5 cm^3) equilibrated with PBS buffer 0.15M, pH 7.0. Elution was carried out using the same buffer to separate CA 15-3 bound to ^{125}I-anti CA 15-3 antibody from unbound (Free) CA 15-3 and ^{125}I-anti CA 15-3 antibody with a flow rate (1 mL per 7 min), and fraction volumes of 1 mL were collected.

3- The radioactivity of each fraction was counted by gamma counter for one minute.

4- Protein concentration was measured at 280 nm.

5- One hundred and twenty microliters of ^{125}I-anti CA 15-3 antibody (0.84 mg. μL^{-1}) was completed to 700 μL with PBS buffer (0.15M, pH7.0), then this volume was injected to the column as mentioned in step2, then steps 2,3 and 4 were repeated.

Calculations

1. Radioactivity (c.p.m) of each eluted fraction was plotted against the fraction number.

2. The absorbance of each eluted fractions was measured at 280nm, and the absorbance was plotted against the fraction number.

3. The percent radioactivity was calculated by dividing the sum of the radioactivity of the fractions under each peak by the sum of radioactivity of all peaks appeared in the profile:

$$\text{Percent radioactivity of each peak} = \frac{\text{Radioactivity per peak (c.p.m)}}{\text{Sum of radioactivity of all peaks (c.p.m.)}} \times 100$$

B) Partially Purified CA15-3 from Premenopausal Malignant Breast Tumors (IDC) and Its Antibody ^{125}I-anti CA 15-3

Reagents

Buffer PBS 0.15 M, pH 7.0 containing 0.02% sodium azid was prepared as described previously in section (2.1.1.3).

Procedure

1. Four hundred and twenty four microliters of partially purified CA 15-3 (0.147 mg. mL^{-1} protein) and incubated with 106 μL of ^{125}I-anti CA 15-3 antibody (0.743 mg.mL^{-1}) in a final volume 700 mL with PBS buffer 0.15M pH 7.0. The tubes were then incubated for 150 min at 15°C.

2. Steps 2,3,4 and 5 in section (5.2.1.3 A) were repeated.

Calculation

The same calculation that mentioned in section (5.2.1.3 A) was used to calculate the radioactivity; protein was measured at 280nm and the percent of radioactivity of each peak was determined.

5.2.2. The UV Spectrum of (125I-Anti CA 15-3 Antibody/CA15-3) Complex from Benign and Malignant Breast Tumors

The gel filtration profile in section (5.2.1.3 A&B) gave two peaks. The fractions under each peak were pooled and the absorption spectrum was scanned in UV Region against the appropriate blank in the reference beam.

5.2.3. The UV. Spectrum of 125I-Anti CA 15-3 Antibody

Half milliliter of ^{125}I-anti CA 15-3 antibody was placed in a 0.25 cm cuvette in the sample beam and the absorption spectrum was measured immediately against an appropriate blank in the reference beam.

5.2.4. The UV Spectrum of Partially Purified CA 15-3

Half milliliter of partially purified CA 15-3 from benign (Fibroadenoma) and malignant (IDC) breast tumors was placed in a 0.25 cm curette in the sample beam and the absorption spectrum was measured immediately against an appropriate blank in the reference beam.

5.2.5.Factors Affecting the Absorption Properties of (125I-Anti CA 15-3 Antibody/CA 15-3) Complex from Benign and Malignant Breast Tumors

5.2.5.1. The pH Effect on the Complex

Reagents

1. KCl-HCl buffer (pH 2) was prepared as follows:

    Solution A: Potassium chloride (0.15M), 1.11825 gm was dissolved in a final volume of 100mL deionized distilled water.

    Solution B: Hydrochloric acid (0.15M).

    The required pH (2.0) was prepared by mixing a volume of solution A with an appropriate amount of solution B to obtain the required pH.

1. Citrate-phosphate buffer at different pH was prepared as follows:

    Solution A: Citric acid (0.15M); 2.8815 gm citric acid dissolved in 100mL deionized distilled water.

    Solution B: Dibasic sodium phosphate (0.15M); 2.1294 gm of Na_2HPO_4 was dissolved in a final volume of 100 mL deionized distilled water.

    Working buffer pH (4 and 6) was prepared by mixing a volume of solution A with an appropriate amount of solution B to obtain the required pH.

2. Phosphate buffer at different pH values was prepared as follows:

    Solution A: Dibasic sodium phosphate (0.15M), 2.1294 gm Na_2HPO_4 was dissolved in a final volume of 100 mL deionized distilled water.

    Solution B: Monobasic sodium phosphate (0.15M), 1.7997 gm NaH_2PO_4 was dissolved in a final volume of 100 ml deionized distilled water.

    Phosphate buffers at different pH rang (7-8) were prepared by mixing a volume of solution A with an appropriate amount of solution B to obtain the required pH.

3. Glycine - NaOH buffer was prepared as follows:

Solution A: Glycin (0.15M); 1.12575gm $C_2H_5NO_2$ was dissolved in a final volume of 100 mL deionized distilled water.

Solution B: Sodium hydroxide (0.15M); 0.6gm NaOH was dissolved in a final volume of 100 mL deionized distilled water.

Working buffer pH (9-11) was prepared by mixing a volume of solution A with an appropriate a mount of solution B to obtain the required pH.

Procedure

Two hundred and fifty microliters of pooled fractions under the first peak that represent ([125I]-anti CA 15-3 antibody/CA 15-3) complex, was completed to 500µl with different buffers at different pH values (4 to 11), then each sample beam and the buffer at the adjusted pH in the reference beam. The absorption spectrum was scanned.

Calculations

The molar absorption coefficient (ε) for ([125I]-anti CA 15-3 antibody/CA 15-3) complex at 278 nm was calculated from Lambert-Beer's law.

5.2.5.2. Effect of Solvent Polarity on UV Spectra of the Complex

The effect of 20% ethanol, and the same amount for ethylene glycol, glycerol, sucrose, urea, dimethyl sulphoxide, dioxane, and polyethylene glycol; on the complex. Two hundred and fifty microliters of complex from benign and malignant breast tumors of pooled fractions under the first peak were completed to 500 µL with phosphate buffer containing any of the following solvent at pH 7.4 in the test cell and the 20% ethanol, ethylene glycol, glycerol, sucrose, urea, dimethyl sulphoxide, dioxane, and polyethylene glycol was adjusted and placed in the reference cell using 0.25 cm cuvette (i.e., the experiment was repeated by using solvents individually).

Calculations

The absorption spectrum of each sample was scanned immediately in the area of (200-350 nm).

5.2.5.3. Spectrophotometric pH Titration on the Complex

A series of complex from benign (Fibroadenoma) and Malignant (IDC) breast tumors (250 µL) were completed to 500 µL with buffer at pH ranging from 8 to 11. The maximum absorbance of each sample was measured at 295 nm; the absorbance of λ_{max} at each pH value was plotted versus the corresponding pH. Other series of complexes isolated from benign (Fibroadenoma) and malignant (IDC) breast tumors (250 µL) were completed to 500 µL with buffer at pH ranging 4 to 8. The maximum absorbance of each sample was measured at 211nm. The absorbance of λ_{max} at each pH value was plotted against the corresponding pH.

5.2.5.4. The Effect of NaCl Concentration on the Thermal Stability of the Complex by UV Spectral Studies

Reagents

Twenty percent ethylene glycol buffur was prepared by dissolving 20mL of ethylene glycol in 80mL of phosphate buffer. NaCl (0.01M) in 20% ethylene glycol was prepared by dissolving 0.05844 gm of NaCl in 100mL of 20% ethylene glycol buffur, while NaCl (0.1M) in 20% ethylene glycol was prepared by dissolving 0.5844 gm of NaCl in 100mL of 20% ethylene glycol buffer.

Procedure

Two hundred and fifty microliters of complex from benign (Fibroadenoma) or malignant (IDC) breast tumors were completed to a final volume 500 μL with 20% ethylene glycol buffer pH7.4 containing 0.01 M NaCl Each mixture was placed in 0.25 cm cuvette in the sample beam and the buffer at the adjusted pH in the reference beam.The absorbtion was measured at the wavelength of (292 and 295 nm) at different temperatures 20, 30, 40, 50, 60, 70°C. The experiment was repeated for each complex with another solution (20% ethylene glycol 0.1 M NaCl), at 295 nm.

Calculations

The absorbance of each complex was plotted against the different temperatures at two wavelengths (292 and 295 nm).

5.2.5.5. Effect of Urea, KCl and (Urea, KCl) Mixture on the Spectrum of the Complex

Reagent

1. Eight molar of urea was prepared by dissolving 24.02gm of Urea in a final volume of 50 mL of PBS buffer at pH 7.4.
2. KCl (0.03 M) was prepared by dissolving 0.2737gm of the salt in a final volume of 50mL of corresponding buffer.
3. PB buffer solution was prepared as described in section (5.2.5.1).

Procedure

Two hundred and fifty microliters of complex isolated from benign (fibroadenoma) and malignant (IDC) were pipetted in a set of three tubes. The volume was completed to 500 μL with PBS buffer at pH 7.4 contains (0.03 KCl, 8 M urea and mixture 1:1 of both 0.03 KCl and 8M Urea) respectively, then each sample was placed in 0.25cm cuvette in the sample beam and the buffer at the same pH in the presence of the same salt in the reference beam.

Calculations

The absorption spectrum of each sample was scanned immediately in the area of (200-350 nm).

5.3. Results and Discussion

Protein UV light maximum absorption is at approximately 280nm, caused by tryptophan, tyrosine and (to a lesser extent) phenylalanine residues, and at lower wavelength (215-230 nm) due to polypeptide chain backbone. Absorbance at 280 nm varies for each protein. The absorbance at lower wavelengths is directly related to the amount of polypeptide material and is usually considerably more sensitive than at 280nm.however,many buffers and other molecules also absorb at these lower wavelengths (phosphate and tris buffers are acceptable but the preservative sodium azide absorbs strongly).

Absorbance at 215-230 nm is useful for monitoring peptides that may not contain tryptophan or tyrosine [205].

Gel Filtration Technique for Separation of Free and Bound 125I-Anti CA15-3 Antibody

Figure (5-1) and (5-2) show the results of gel filtration technique to separate ^{125}I-anti CA 15-3 antibody bound to partially purified CA 15-3 from benign (Fibroadenoma) and malignant (IDC) breast tumors respectively. The profile of separation revealed two peaks. The first peak represents (^{125}I-anti CA 15-3 antibody/CA 15-3) complex, the second peak represents the unbound (Free) ^{125}I-anti CA 15-3 antibody. Figure (5-3) show the gel filtration profile of ^{125}I-anti CA 15-3 antibody, the results revealed only one peak in the same position of the second peak of figures (5-1) and (5-2), which represent the unbound ^{125}I-anti CA 15-3 antibody. The percent of ^{125}I-anti CA 15-3 antibody/ CA 15-3) complex was 49.74% in benign (Fibroadenoma) breast tumors patients, while the percent of complex was 56.20% in malignant breast tumors patients (IDC). On the other hand the percent of ^{125}I-anti CA 15-3 antibody was 34.40% in benign breast tumors (Fibroadenoma) and 31.42% in malignant breast tumors (IDC). This is because the epitope of CA 15-3 in malignant breast tumors was higher than in benign breast tumors.

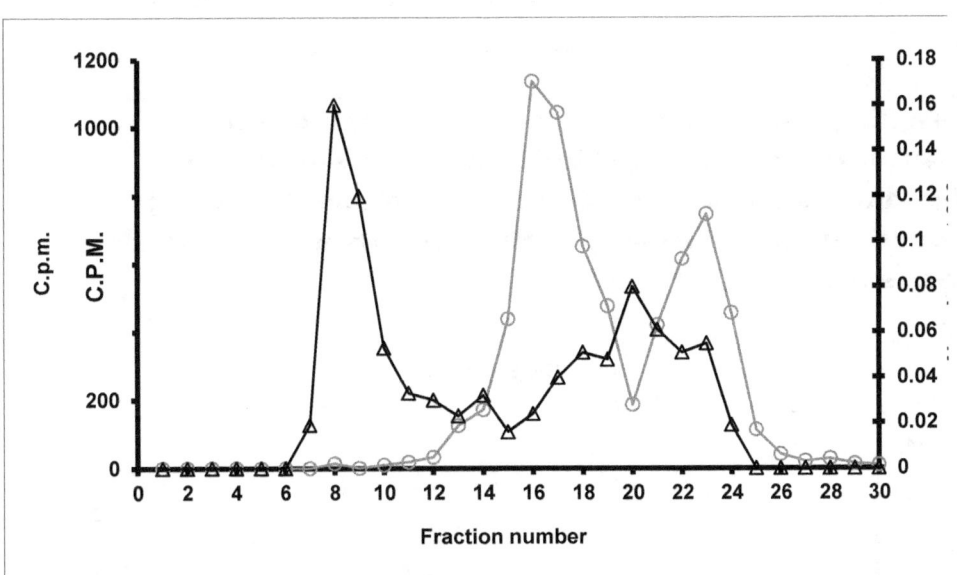

Figure (5.1): The elution profile of the isolated complex (^{125}I-antiCA15-3 antibody/CA15-3) and free antibody in benign breast tumors on Sepharose CL-4B. (O) radioactivity, (△) protein. (All other details are explained in the text).

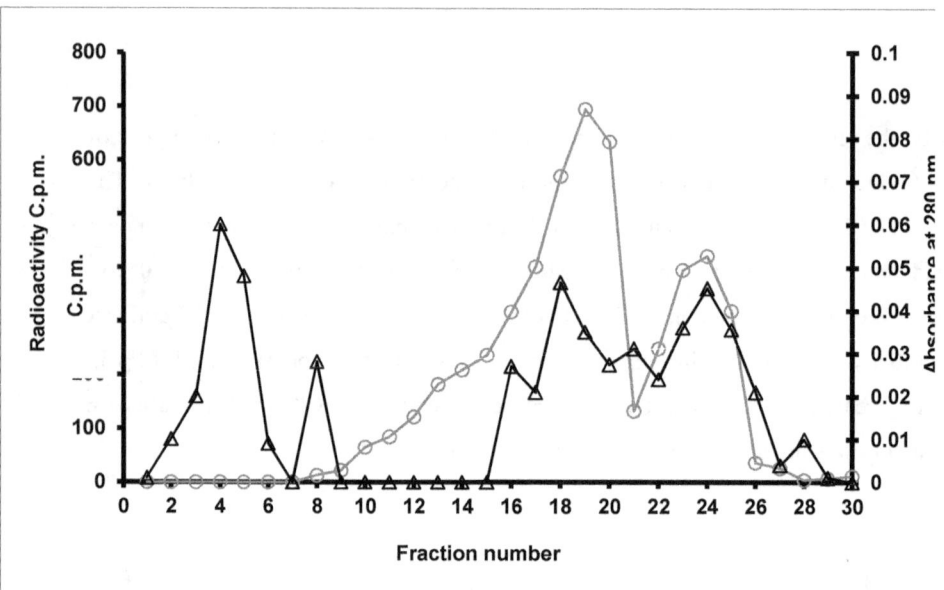

Figure (5.2): The elution profile of the isolated complex (^{125}I-antiCA15-3 antibody/CA15-3) and free antibody in malignant breast tumors (IDC) on Sepharose CL-4B, (○) radioactivity, (△) protein. (All other details are explained in the text).

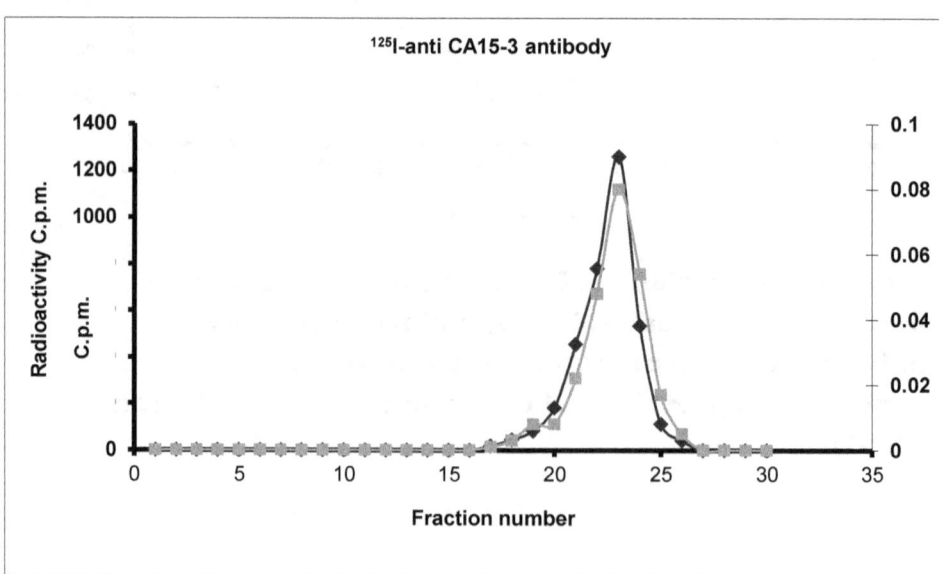

Figure (5.3): The elution profile of the ^{125}I-antiCA15-3 antibody on Sepharose CL-4B, (■) radioactivity, (♦) protein. (All other details are explained in the text).

The UV Spectra of Partially Purified CA 15-3, Anti CA 15-3 Antibody and (125I-Anti CA 15-3 Antibody/CA 15-3) Complex Molecules

The UV spectra of partially purified h-CA 15-3, ^{125}I-anti CA 15-3 antibody and (^{125}I-anti CA 15-3 antibody/CA 15-3) complex were scanned from 200-350 nm to determine the absorption spectra, and the alternation in the UV spectra as a results of their interaction.

The UV Spectrum of Partially Purified CA 15-3

The UV spectra of partially purified h-CA15-3 in benign tumors (Fibroadenoma) and malignant tumors (IDC) at neutral pH shows that the λ_{max} for purified CA15-3 from benign (Fibroadenoma) consisted of two peaks; a large one at 208nm and smaller one at 270nm, while the UV spectra of purified CA 15-3 from malignant tumors (IDC) shows two peaks at 205 and 270nm as shown in table (5.1). Therefore it seemed that each human CA 15-3 has a characteristic spectrum and can be identified by its peaks, the first peak (at 208nm or 205nm) such results could be due to the amide group in polypeptide bond of h-CA 15-3 molecule with contribution of the histidyl residues [207], while the second peak (at 270) is assigned to the side chain chromophore of phenylalanine or tryptophyl residues [208].

The UV spectrum of 125I-Anti CA 15-3 Antibody

The UV spectrum of ^{125}I-anti CA 15-3 antibody at neutral pH shows that the λ max consisted one peak at 203.6nm, which is assigned to the amide groups in the polypeptide bond [207], with contribution of hisidyl residues [209] as shown in table (5.1) .

The UV spectrum of (125I-Anti CA 15-3 Antibody/CA 15-3) Complex

The UV spectra of partially purified CA 15-3 extracted from benign (Fibroadenoma) and malignant (IDC) bound to ^{125}I-anti CA 15-3 antibody at neutral pH show that the λ_{max} is consisted of two peaks at (203.4nm and 274nm) in benign complex, while the λ_{max} is consisted of two peaks at (204.2 nm and 278nm) in malignant complex as shown in table (5-1). The first peak at (274nm or 278nm) is assigned to tyrosyl residues[208], it is very weak band and it seems that the tyrosyl residues in the benign or malignant complexes is located on the surface of protein complex.

The strong absorption of the second peaks (at 203.4 or 204.2nm) arises form electronic transition in the peptide backbone itself and is therefore sensitive to backbone conformation [207].

Table (5-1): The λ_{max} valves of (^{125}I-anti CA 15-3 antibody/CA 15-3) complex, partially purified CA 15-3 and unbound (Free) ^{125}I-anti CA 15-3 antibody in both cases benign and malignant breast tumors. (All other details are explained in the test).

No.	Fractions	Benign λmax (nm)	Malignant λmax (nm)
1	CA 15-3 partially purified	208, 270	205, 270
2	^{125}I-anti CA 15-3 antibody	203.6	203.6
3	^{125}I-anti CA15-3 antibody/ CA15-3) complex	203.4, 274	204.2, 278

Factors Affecting the Absorption Properties of (125I-Anti CA15-3 Antibody/ CA15-3) Complex from Benign and Malignant Breast Tumors

The Effect of pH on the Complex

The pH of the solvent determines the ionization state of ionizable chromophores in the protein molecule [208]. The UV spectrum of isolated (^{125}I-anti CA 15-3 antibody/CA15-3) complex from benign (Fibroadenoma) and malignant (IDC) breast tumors was determined at different pH (2, 4, 6, 7, 7.4, 8, 9, 10, and 11). Table (5-2) shows the effect of different pH on both complexes. At an acidic pH 2 and neutral pH (7,7.4) the both complexes benign (Fibroadenoma) and malignant (IDC) have one maximum wavelength near 200 nm as compare to UV spectrum of h-CA 15-3 and the ^{125}I anti CA15-3 antibody.

The λ_{max} of CA 15-3 (270 and 208 nm) in benign (Fibroadenoma) and its antibody λ_{max} (203.6) disappeared. The λ_{max} of CA 15-3 (270 and 205 nm) in malignant (IDC) and its antibody λ_{max} (203.6) also disappeared. The absorption near

200 nm is characteristic of the amide group in the polypeptide bond of the complex [209]. The blue shift is due to the increasing of hydrogen bond formed in the presence of highly positively charged state [210]. The disappearance of λ_{max} 280 nm of tyrosine and phenylalanine due to conformational changes and chromophore in native complex were buried in the interior of their complexes [211,212]. Protein shows a strong absorption in range (180-225 nm), absorption at such wavelength arises from electronic transition in the polypeptide backbone itself and is therefore sensitive to back bone conformation [207].

At pH (4, 6, 9, 10, and 11) no band was observed and all peaks disappeared. The disappearance of the λ_{max} at these pH's may be due to conformational changes of the protein complex.

Table (5-2): The effect of different pH on λ_{max} values of (^{125}I-anti CA 15-3 antibody/CA 15-3) complex. (All other details are explained in the text).

pH	λ_{max} (nm)	
	(^{125}I-anti CA 15-3 antibody/CA15-3) benign complex	(^{125}I-anti CA 15-3 antibody/CA 15-3) malignant complex
2	200	200
4	-	-
6	-	-
7	200	200
7.4	200	200
8	200	200
9	-	-
10	-	-
11	-	-

Effect of Solvent Polarity on UV Spectra of the Complex

The immediate environment of a chromophore affects its absorption. The determination of whether an amino acid is internal or external by measuring the spectra of protein in a polar and non-polar solvent is called the solvent perturbation method [208]. In fact, proteins are rarely studied in completely non-polar solvents because most proteins are either insoluble or denatured in these solvents. However, significant solvent effects can be induced by use of a mixture of water and substance of a reduced polarity such as ethanol, ethylene glycol, polyethelyne glycol, sucrose, dioxane and dimethyl sulfoxide (DMSO)[208]. Several spectra changes were obtained in the precence of these perturbants, like the alteration of λ_{max} positions and intensities of protein spectrum and the appearance of new chromophores on the surface of the complex. These chromophores on the region of the protein disappeared in the absence of the solvent. One of the main assumptions of the solvent perturbation technique is that solvent alters the peak positions and intensities by altering the energy and probably of electronic transitions. Other considerations include the following [213,214]:

a. Polarization effect

b. Change in permanent dipole moment during excitation, which will tend to produce either a short wave or a long wave shift depending on the nature of the electronic transition and wheather the solute is a hydrogen donor or hydrogen acceptor [215].

The effects of different solvents on the (^{125}I-anti CA 15-3 antibody /CA 15-3) complex from benign (Fibroadenoma) and malignant (IDC) breast tumors at pH 7.4 were investigated. The data obtained are illustrated in table (5-3). It was found that one λ_{max} specific for the amide groups of polypeptide bond at pH 7.4, this shift toward the shorter wavelength is due to the n-π* transitions in the presence of 20% ethanol, ethylene glycol and glycerol. In the presence of polyethylene glycol there was a significant red shift in the λ_{max} (204 nm) in benign (Fibroadenoma) complex and λ_{max} (205nm) in malignant (IDC) complex. When 20% Dioxane was used there were a significant red shift in the λ_{max} (220nm) of the amide bond at pH 7.4, which assigned to tyrosyl residue. The value of λ_{max} is for n-π* transitions which occur at longer wavelength because the nonbonded electrons in the anion are available for interaction with the π electron system of the ring [215], while in the presence of 20% sucrose the complex has a slight blue shift and show λ_{max} at 202 nm and 201nm in both benign and malignant complexes. Finally the effect of 20% DMSO on the complex, show that the amide bands at pH 7.4 were disappeared, this may be due to the denaturation of protein complex in presence of 20% DMSO.

The application of spectrophotometric solvent perturbation on the complex is to determine the location of tyrosyl residues, whether they are buried and inaccessible or exposed and accessible to the solvent approach [216]. Laskowski [201] has listed the major assumptions of solvent perturbation experiments. There are: (1) buried chromophors are unperturbed, that is only the groups located on the surface or near the surface of the protein should experience the perturbing effects of the solvent; groups buried in the interior of the protein, not accessible to the solvent; which should not be affected and consequently could not contribute to the overall spectral shift observed. (2) No conformational changes take place upon addition of perturbant, and (3) the solvation layer around the chromophore contains the same concentration of perturbation experiments when employed at convenient concentrations (often 20%), do not appear to produce conformational changes in most protein studied under reasonable conditions of pH, ionic strength, and temperature. This concentration is large enough to cause measurable shifts in the spectra of chromophoric residues. Conformational changes can be expected if perturbation is carried out under conditions in which the protein structure has marginal stability (low-or high pH for many protein)[201,216]. Chromophore may not completely bury. It has been, distinguished between chromophores in crevices and chromophores that are partially buried [201]. The former are observed to be fully perturbed by perturbant solvent smaller than a certain critical size (e.g ethanol), but not by larger perturbants (e.g polyethylene glycol). The degree of exposure is thus determined only by the size of perturbant molecule (solvent) or by the size of the crevice in which the chromophore is located [216]. Partially buried chromophores on the other hand, show a degree of exposure that depends on the nature of the perturbant, rather than on its size. The observed degree of exposure decrease in the order:

Sucrose $\geq$ Glycerol $\geq$ Ethyleneglycol $\geq$ Methanol, Ethanol $>$ Polyethylene glycol $\geq$ Dimethyl sulfoxide.

The first perturbant in this series modify the solvent nonspecifically, while the later ones in the series may specifically interact with chromophore[216].

When comparing the effects of the six solvents used, ethanol, ethleneglycol, glycerol, dioxane, polyethylene glycol and dimethylsulfoxide at pH 7.4 on the UV spectrum, especially on the shift of λ_{max}, which is due to tyrosyl residues. It seems that the maximum effect was observed in the presence of 20% dioxane as perturbant solvent, where there was a shift in the λ_{max} about 16nm, while minimum effect was observed in the presence of 20% polyethylene glycol where the λ_{max} remained unchanged. Since the change in the λ_{max} of tyrosyl residues does not depend on the size of the perturbant solvent, the tyrosyl residues showing the changes in λ_{max} ; absorbance must be partially buried.

Table (5-3): The effect of 20% of ethanol, ethyleneglycol, glycerol, polyethylene glycol, sucrose, dioxane, DMSO and on the λ_{max} of (^{125}I- Anti CA 15-3 Antibody/ CA15-3) complex at pH 7.4. (All other details are explained in the text).

Solvent of 20% of	λ_{max} (nm)	
	(^{125}I-anti CA 15-3 antibody/ CA15-3) benign complex	(^{125}I-anti CA 15-3 antibody/ CA 15-3) malignant complex
Ethanol	Near 200	200
Ethylene glycol	Near 200	Near 200
Glycerol	Near 200	Near 200
Polyethylene glycol	204	205
Sucrose	202	201
Dioxane	220	220
DMSO	-	-

Spectrophotometric pH Titration of the Complex from Benign and Malignant Breast Tumors

To study (^{125}I- anti CA 15-3 antibody/ CA15-3) complex structure, this requires the determination of pk_a values for proton dissiociation from ionizable amino acid side chains, because these values give an indication of the location of amino acid in the protein. This can often be done spectrophotmetrically because dissociation often changes the spectrum of one of the chromopores (tyrosyl)[208]. For proteins this usually amounts to the titration of the phenolic groups of tyrosine residues. By the measurement of the absorption at 295 nm (λ_{max} for the ionized form of tyrosine), or observation of histidine dissociation by measurment at 211nm.

The titration curves of (^{125}I- anti CA 15-3 antibody/ CA15-3) complex from benign (Fibroadenoma) and malignant (IDC) for both histidyl and tyrosyl residues are illustrated in figure (5-4 A&B) respectively. Figure (5-4) shows that the pk_a for histidine is (6.69) for (^{125}I- anti CA 15-3 antibody/ CA15-3) complex from benign breast tumors, while the pk_a for histidine is (6.65) for (^{125}I- Anti CA 15-3 Antibody/ CA15-3) complex from malignant (IDC) breast tumors. From the same curve it could be concluded that about (41.43%) histidyl residues are located on the surface of the protein complex [217] of benign (Fibroadenoma), while about (44.29%) histidyl residues are located on the surface of the protein complex from malignant (IDC). The other residues are buried interior the benign and malignant complex. Figure (5-4 B) shows that the pk_a value of the benign complex of tyrosyl residues is (8.9) and it's about (40%) at tyrosine residues are internal and a large arise in the absorbance at very high pH was observed. While in the malignant (IDC) complex the pk_a value of tyrosyl is (8.4) and it's about (50%) this indicates that the internal tyrosines have become exposed to the solvent, which is the protein complexes in folded (become denatured)[217].

The two curves also illustrated the low content of histidine compared to the high content of tyrosine in the benign and malignant complex.

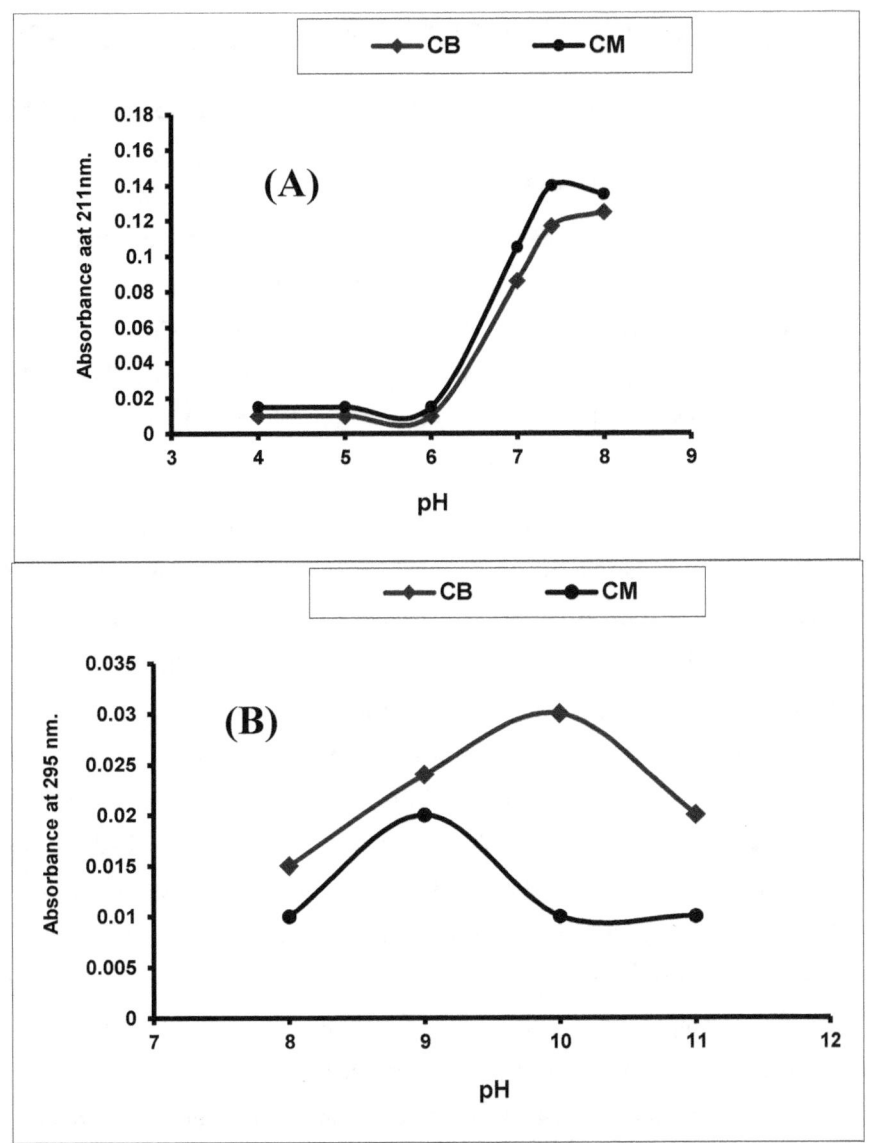

Figure (5.4): Spectrophotometric pH titration of ^{125}I-anti CA15-3 antibody/CA15-3 complex from benign and malignant breast tumors:

(A) for histidine, (B) for tyrosine.

(CB): Complex of benign breast tumors, (CM): Complex of malignant breast tumors. (All other details are explained in the text).

The Effect of NaCl Concentration on the Thermal Stability of the Complex by UV Spectral Studies

The effect of different concentrations of NaCl on the thermal stability of the protein complex isolated from benign and malignant breast tumors was examined in this experiment. The values of absorbance at λ_{max} (292, 295nm) for tryptophyl and tyrosyl residues respectively, in two different concentrations of NaCl 0.01 M and 0.1 M in 20% ethylene glycol buffer are shown in figure (5.5 A&B) and (5.6 A&B). The λ_{max} was used to examine if the protein contains internal tryptophans and tyrosines .

As shown in figure (5.5 A&B), the absorbance of both tryptophane and tyrosine reach higher absorbance at 60°C, in the presence of 0.01 M NaCl in benign and malignant complex. The increment in the absorbance of both tryptophyl and tyrosyl residues with increasing temperature could be due to that buried chromophores becomes exposed to the solvent during thermal denaturation [209].

Figure (5.6 A) shown the absorbance of tyrosin reach higher absorbance at 70°C in the presence of 0.1 M NaCl in benign and malignant complex. On the other hand figure (5.6 B) shown the absorbance of tryptophane reach higher absorbance at 60°C and 30°C in benign and malignant breast tumors complexes in presence of 0.1 M NaCl respectively. Which means that the complexes were very stable at 70°C in presence of higher concentration of NaCl, 70°C was needed for unfolding benign and malignant complex at λ_{max} 292 nm and benign complex was more stable at 60°C in presence of 0.1 M NaCl,while the temperature is decreased to 30 °C in the presence of 0.1M NaCl at λ_{max} 295. This is due to conformational changes required more energy 70°C in presence of 0.1 M NaCl than in 0.01 M NaCl.

The decreased absorbance in presence of 0.1 M NaCl as compared with that in 0.01 M NaCl could be due to salt concentration. Each protein in solution containing salts will collect around it a counter ion atmosphere enriched in oppositely charged small ion (chloride ion, sodium ion) and such a cloud of ions will tend to screen the protein, the more effective electrostatic screening will be, and decrement in the absorption intensity will be observed [207].

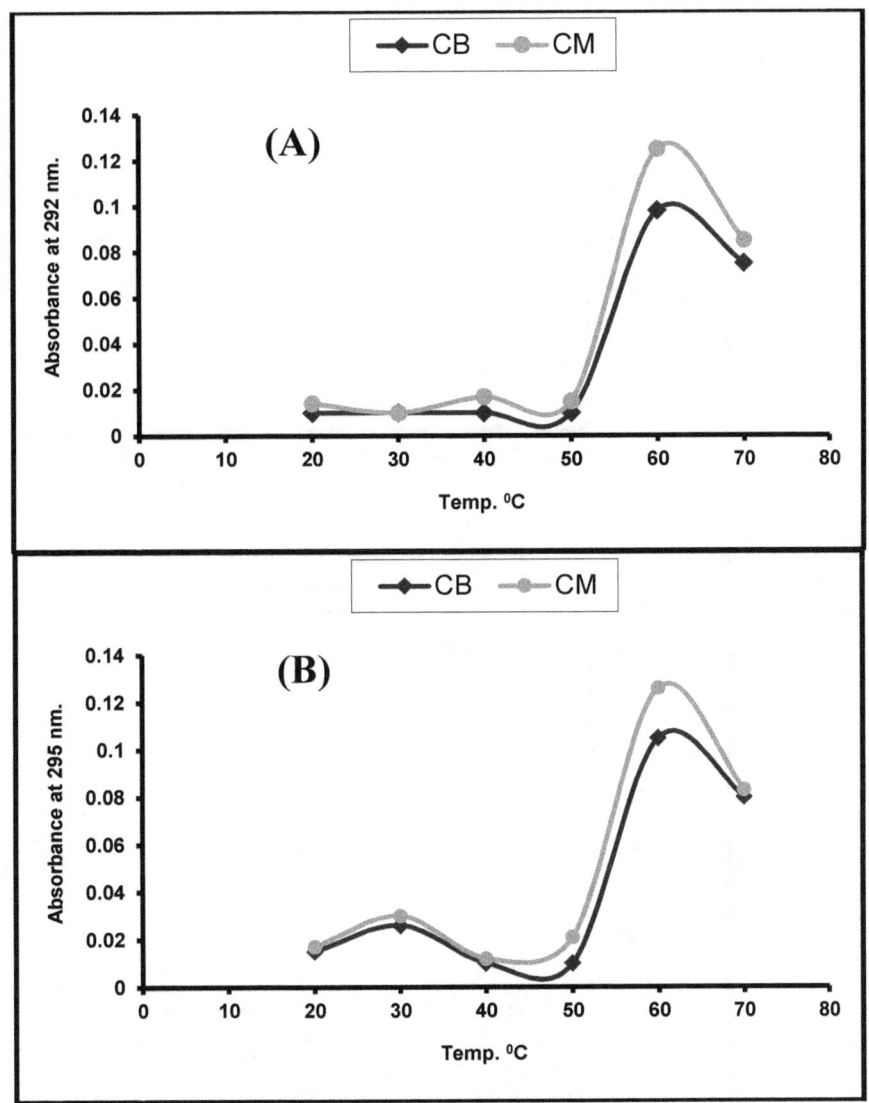

**Figure (5.5): Thermal stability curve for benign and
malignant: (A) at λ_{max} 292 in the presence
of 0.01 M NaCl, (B) at λ_{max} 295 in the
presence of 0.01 M NaCl.
(CB): Complex of benign breast tumors,
(CM): Complex of malignant breast tumors.
(All other details are explained in the text).**

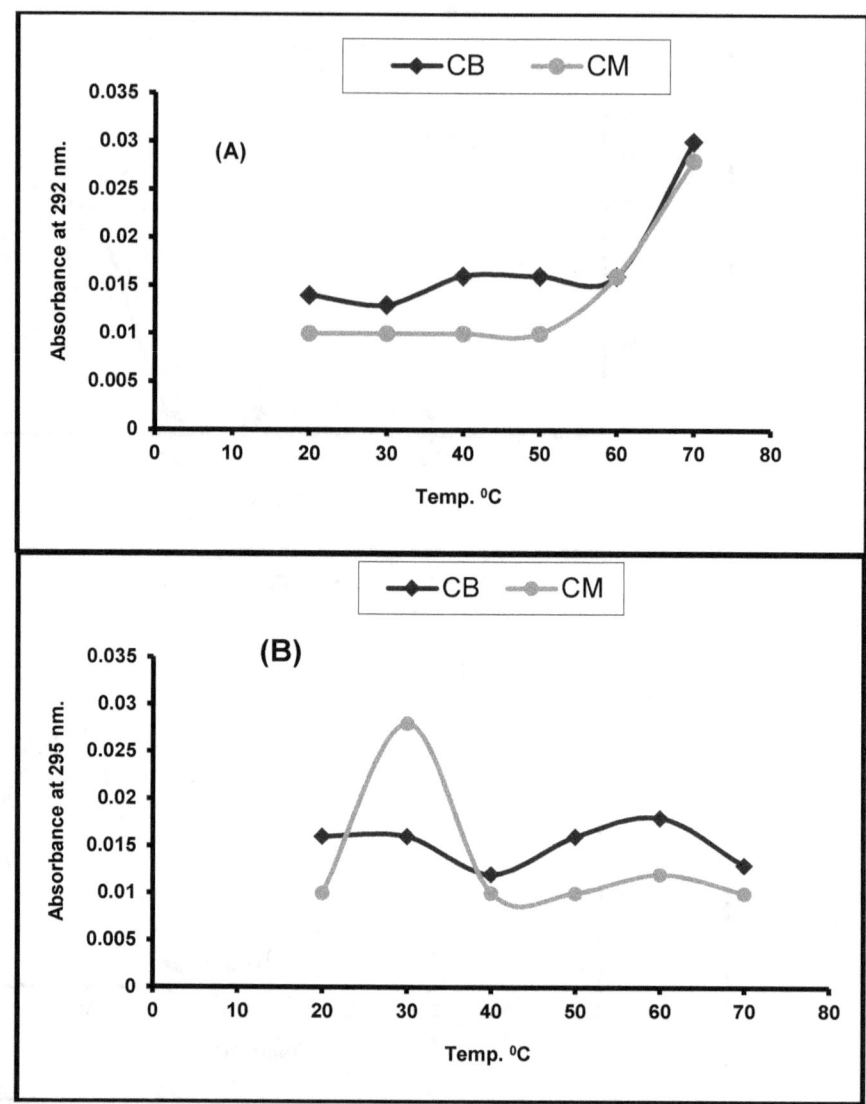

**Figure (5.6): Thermal stability curve for benign and
malignant: (A) at λ_{max} 292 in the presence
of 0. 1 M NaCl, (B) at λ_{max} 295 in the
presence of 0.1 M NaCl.
(CB): Complex of benign breast tumors,
(CM): Complex of malignant breast tumors.
(All other details are explained in the text).**

Effect of Urea, KCl and (Urea, KCl) Mixture on the Spectrum of the Complex

The effect of 8 M urea, 0.03 M KCl and a mix of 1:1 of 8 M urea and 0.03 M KCl on the λ_{max} of the benign (fibroadenoma) and malignant (IDC) complexes, were examined. The values of λ_{max} are illustrated in table (5-4). When table (5-4) is compared with table (5-1), it seems that the presence of 8 M urea at pH 7.4, there was a red shift of the λ_{max1} of polypeptide bond from 200 to 227.4 nm in benign complex and a red shift of λ_{max1} from 200 to 226 nm in malignant complex respectively. While λ_{max2} of aromatic amino acid i.e., tyrosine residues in both complexes was disappeared. The red shift is due to intramolecular hydrogen bonding between the oxygen of the amide group and the solvent [218].

When 0.03 M KCl was used, there was no alternation in the position of the λ_{max2} of the tyrosyl at pH 7.4 in both benign and malignant complexes. There was a slight blue shift (3-4nm) in the λ_{max1} of the polypeptide bond in the benign and malignant complex spectra respectively. On the other hand the λ_{max} of the aromatic ring of tyrosyl residues at (274 or 278nm) disappeared. Such blue shift can arise by introducing positive (K^+) or negative (Cl^-) charges near the chromophore (the amid group), which might interact with π-electron system of the amide group [201].

When 8 M urea was mixed with 0.03 M KCl there was significant red shift in λ_{max} (203.4 and 204.2nm) to λ_{max} (221.4 and 219.4nm) in both benign and malignant complexes. The same shift was observed when 8 M urea was used alone with each benign and malignant complexes, this mean that the red shift due to the effect of urea, but not to 0.03 M KCl. On the other hand, there was no alternation in positions of the λ_{max} of the tyrosyl residues near 278nm.As was seen,the changes in absorption were near 230 nm and near 280 nm. This was also observed by Glazer who that solvent perturbation or denaturation of protein poduces may changes in absorption near 230 nm and 280 nm. Some of this change in absorption may be produced by change in the n-$\pi*$ absorption of poly peptide bond in protein either because of a change in their geometrical arrangement, or because of an environment changes [219].

Table (5-4): The effect of 8M urea, 0.03M KCl and mixture (urea+KCl) on the λ_{max} of the complex UV

spectrum at pH 7.4. (All other details are explained in the text).

Solvent	λ$_{max}$ (nm)	
	(^{125}I-anti CA 15-3 antibody/ CA 15-3) Benign Complex	(^{125}I-anti CA 15-3 antibody/ CA 15-3) Malignant Complex
Urea 8M	227.4	226
KCI 0.03M	200	200
Urea+KCI mixture 1:1	221.4 278.6	219.4 278

chapter six

Immunoradiometric assay

Asolid-phase Immunoradiomertric Assay sandwich technique (IRMA) was used for the determination of the carbohydrate antigen 19-9 (CA19-9) defined by a monoclonal antibody ^{125}I-anti CA19-9. The antibody ^{125}I-anti CA19-9 reacts with CA19-9 found at low concentrations in sera of healthy women but increased slightly in sera of patients with breast cancer.

The factors affecting the binding of ^{125}I-anti CA19-9 antibody with CA19-9 in the breast tumor homogenate (benign and pre-and post-menopausal malignant) were determined. The results revealed that 100, 75 and 75 µg protein was the most appropriate amount of protein used in each incubation at pH 7.8, 8.0 and 7.0 respectively, with 0.0565 mg. mL^{-1} of ^{125}I-anti CA19-9 antibody for 4,1 and 6 h incubation time at optimum temperatures 25, 37 and 45 °C respectively. The use of 0.01 M sodium halides and 0.025 M of divalent salts were shown to cause different effects on the binding in the three groups.

The recovery of the method was calculated and found to be 99% , 98% and 95% for binding CA19-9 present in (benign and pre-and post-menopausal malignant) breast tumor homogenates respectively.

Introduction

The Carbohydrate antigen 19-9 (CA19-9) (Koprowski etal.,1979) [118], is specific carbohydrate fraction of a circulating antigen found in sera of normal adults (Koprowski etal.,1981) [220], has sialyl Lewis^a structure and is present in individually expressing the Lewis^a and /or Lewis^b blood group antigen [114]. CA19-9 is identified as a glycolipid- that is , sialylated lacto-N-fucopentose II ganglioside [221]. In serum, it exists as a mucin , a high molecular mass (200-1000 KD) glycoprotein complex [54]. In Normal tissues, sialyl Lewis^a antigen is present in ductal epithelium of breast, kidney, salivary gland, and sweatglands[115-117].

CA19-9 is measured with a double monoclonal immuno-radiometric assay [178]. Another techniques used for the detection of CA19-9 in tissues and sera were performed by an immunoperoxidase assay [126] and by radioimmunoassay [125] of samples from patients, and enzyme immunoassay [124] for quantitative determination of CA19-9 in human serum. The upper limit of normal value 37.0 U.mL^{-1} [121,222]. The abnormal expression of the sialyl Lewis ^{a} is closely correlated with various forms of cancer including pancreatic cancer [223-225], gall bladder [226] and bile duct [227] cancer.

A monoclonal antibody CA19-9 against sialyl Lewis [a] is a popular diagnostic agent for these tumors. The antibody is useless for cancer diagnosis when a patient is lacking the enzyme for the synthesis of sialyl Lewis [a]. In Japan, about 5-10% of the population lacks this enzyme leading to false negative results [228]. CA19-9 represents the most important and basic carbohydrate tumor marker. The immunohistologic distribution of CA19-9 in tissues is consistent with the quantitative determination of higher CA19-9 concentrations in cancer than in normal of tissues [126,229]. Recently reports indicates that serum CA19-9 level is frequently elevated in the serum subjects with pancreatic (80%), hepatobiliary (67%), gastric (40-50%), hepatocellular (30-50%), colorectal (30%) and breast (15%) cancer [51].

Research studies demonstrate that serum CA19-9 values may have utility in monitoring subjects with the above-mentioned diagnosed malignancies [230-232]. A declining CA19-9 value may be indicative of a favorable prognosis and good response to treatment [233]. Therefore, the development of immunoradiometric assay was planned to carry out the determination of the optimum conditions of ^{125}I-anti CA19-9 antibody.

Materials and Methods

6.1. Materials

6.1.1. Chemicals

All chemical and reagents mentioned in the section (2.1.1) were used in the experiments of this chapter; other reagents used were indicated in each experiment.

6.1.2 Instruments

All instruments described in section (2.1.2) all were used in the experiments of this chapter.

6.1.3 Patients and Blood Samples

Thirty breast patients and specimens mentioned in section (2.1.3) were used in this chapter, classified to three group of patients, one group with benign and two groups with malignant breast tumors. The fourth group is a healthy women used as control.

- **Group I:** Consisted of 10 patients with benign (Fibroadenoma) breast tumors.

- **Group II:** Consisted of 10 premenopausal patients with breast cancer (IDC).

- **Group III:** Consisted of 10 postmenpausal patients with breast cancer (IDC).

- **Group IV:** Consisted of 10 normal healthy subjects.

Blood samples were prepared as described in section (2.1.4). PBS buffer was prepared as described in section (2.1.6), while homogenization of breast tumor tissues was carried out as described in section (2.1.7). Statistical analysis was determined by student's t-test as mentioned in section (2.1.8).

6.2. Methods

6.2.1. Determination of CA19-9 Levels in Sera of Patients with Benign and Malignant Breast Tumors

Reagents

The reagents IRMA-ELSA CA19-9 Kit was provided from CIS-bio international ORIS Group/France.

1. Anti CA19-9 monoclonal antibody coated on the ELSA fixed in the bottom of the tube.
2. Anti ^{125}I-CA19-9 monoclonal antibody, radioactivity content < 10 μCi (<370 KBq)
3. Six standard ready for use, Human serum, Human CA19-9 in sodium azide (0,14,30,66,130 and 255 U.mL^{-1}).
4. Diluent (0.0 U.mL^{-1}), human serum in sodium azide.

5. Control (35 U.mL^{-1}), human serum, human CA19-9 in sodium azide. Patients sera and control were used without dilution in this assay.

Procedure

The assay protocol is described in table (6-1).

Table (6.1): IRMA protocol of serum CA19-9 (U.mL^{-1}).

	CA19-9 (U.mL^{-1})						Control		Unknown Samples	
	0	14	30	66	130	255	Level I	Level II	1	2 etc.
Coated tube no.	1,2	3,4	5,6	7,8	9,10	11,12	13,14	15,16	17,18	19,20
Standards (µL)	←————————————— 100 µL —————————————→									
Control serum or samples (µL)	←————————————— 100 µL —————————————→									
Buffer (µL)	←————————————— 200 µL —————————————→									
	Incubation for 3 h. at 37 °C in water bath									
	The solution was aspirated, and washed the tubes 3 times with 3 mL distilled water									
^{125}I-anti CA19-9 (µL)	←————————————— 300 µL —————————————→									
	All tubes were mixed gently with vortex-type mixer and									
	Incubated for 3 hrs. at room temperature (18-25 oC)									
	The solution was aspirated, the tubes were washed 3 times with 3 mL distilled water									
	The remaining bound radioactivity was measured with gamma counter.									

Calculations

1. The mean net count for each group of tubes was counted in gamma counter for 1 min, represents the bound c.p.m.

2. The standard curve was constructed by plotting counts per min. (Y-axis) versus concentration of CA19-9 standard (X-axis) figure (6.1). Then the points were connected with straight-line segments.

6.2.2. Preliminary Test of the Binding of CA19-9 in Breast Tumor Tissues with 125I-Anti CA19-9 Antibody in Breast tumors Homogenates

Reagents

Phosphate buffered saline pH 7.2 was prepared as described in section (2.1.6).

Procedure

The pellet and the cytosol fractions were obtained from the supernatant of breast homogenate were centrifuged at 4000 r.p.m. In order to detect CA19-9, 20 μL of crude cytosol fraction having 1100 μg protein were incubated with 60 μL (0.1356 mg.mL^{-1}) of ^{125}I-anti CA19-9 antibody. The volume of mixture was completed to 500 μL with PBS buffer pH 7.2, and then incubated at 37 °C for 3 hrs. The assay tubes were centrifuged at 4000 r.p.m. for 45 min. at 45 °C. The supernatant was discarded, the rims at tube were swabbed with cotton piece, then the complex formed was counted in gamma counter for 1 min. Pellet

CA19-9 were determined by dissolving the sediment in PBS buffer pH 7.2 with ratio 1:5 (weight: volume), then 20 μL of supernatant fraction of pellet breast homogenate having 800 μg protein, was added to 60 μL (0.1356 mg.mL^{-1}) of ^{125}I-anti CA19-9 antibody. The same steps mentioned above were followed to determine the radioactivity of the complex formed. For total radioactivity two additional tubes with 60-μL of ^{125}I-anti CA19-9 antibody were counted in gamma counter.

Calculations

1. The counted radioactivity in each tube (expressed in c.p.m.) represents the bound fraction (B); (i.e., ^{125}I-anti CA19-9 antibody/CA19-9 complex).
2. The counted radioactivity in the tubes counting ^{125}I-anti CA19-9 antibody only represents the total radioactivity (T).
3. The (B/T) % ratio for each tube was calculated as follows:

$$(B/T)\% = \frac{Sample\,counts(B)}{Total\,counts(T)} \times 100$$

6.2.3.Factors Effecting of 125I-Anti CA19-9 Antibody Binding to CA19-9 in Breast Tumors Homogenates

6.2.3.1. Effect of Protein Concentration on the Binding

Reagents

All reagents prepared is described in section (2.1.6) and (2.2.3.1).

Procedure

Sixty microliters (0.1356 mg.mL^{-1} protein) of ^{125}I-anti CA19-9 antibody were added to 20 μL of cytosolic fraction of benign (Fibroadenoma) and malignant (premenopausal IDC and postmenopausal IDC) breast tumors respectively, containing increasing amounts of protein (50, 75, 100, 150, 200 and 250 μg.mL^{-1}) and were completed to a final volume of 500 μL with 0.15 M PBS pH 7.2. The assay tubes were incubated for 3 hrs. at 37 °C. At the end of incubation, the assay tubes were centrifuged at 4000 r.p.m. for 45 min. at

4 °C. The supernatant was decanted; the rims at the tube were swabbed with cotton piece. The radioactivity of the complex formation was counted using gamma counter.

Calculations

1. The (B/T) % values were determined as in section (6.2.2).
2. Values of (B/T) % were plotted against their corresponding amount of protein of the breast tumor homogenate.

6.2.3.2. Effect of 125I-Anti CA19-9 Antibody Concentration on the Binding

Reagents

All reagents prepared is described in section (2.1.6) and (2.2.3.1).

Procedure

Sixty microliters of increasing amounts (0.0226, 0.0452, 0.0565, 0.113, 0.1356, 0.226 mg.mL^{-1}) of ^{125}I-anti CA19-9 antibody were added to 20 μL of crude cytosolic fraction (100, 75 and 75 μg protein) for benign (fibroadenoma) and malignant (premenopausal IDC and postmenopausal IDC) respectively, completed to a final volume 500 µL with 0.15 M PBS pH 7.2. After incubation for 3 hrs at 37 °C the bound CA19-9 was determined as mentioned in section (6.2.2).

Calculations

1. **The (B/T) % values were determined as in section (6.2.2).**
2. Values of (B/T) % were plotted versus the concentrations of ^{125}I-anti CA19-9 included.

6.2.3.3. Effect of pH on the Binding

Reagents

All reagents prepared is described in section (2.1.6) and (2.2.3.1).

Procedure

Twenty microlites (100, 75 and 75 µg protein) of cytosolic fraction (fibroadenoma, premenopausal IDC and postmenopausal IDC respectively) were added to 25 µL (0.0565 mg.mL^{-1}) of ^{125}I-anti CA19-9 antibody respectively.The volume of the mixture was completed with PBS buffer of different pH (6.8, 7.0, 7.2, 7.4, 7.6, 7.8, 8 and 8.2) to a final volume 500 μL. After incubation for 3hrs at 37 °C, the bound CA19-9 was determined as mentioned in section (6.2.2).

Calculations

1. The (B/T) % values were determined as in section (6.2.2).
2. Values of (B/T) % were plotted versus the corresponding pH.

6.2.3.4. Effect of Temperature on the Binding

Reagents

All reagents prepared is described in section (2.1.6) and (2.2.3.1).

Procedure

Twenty microliters (100, 75 and 75 μg protein) of cytosolic fraction (Fibroadenoma , premenopausal IDC and postmenopausal IDC) were added to 25 μL (0.0565 mg.mL^{-1}) of ^{125}I-anti CA19-9 antibody respectively. The volume of mixture was completed to a final volume 500 μL with PBS buffer at pH 7.8 for fibroadenoma , pH 8.0 for premenopausal (IDC) and pH 7.0 for postmenopausal (IDC). The experiment was carried out at (5, 15, 25, 37 and 45°C) for 3hrs. After incubation the bound CA19-9 was determined as mentioned in section (6.2.2).

Calculations

1.The (B/T) % values were determined as in section (6.2.2).

2.Values of (B/T) % were plotted versus the temperature.

6.2.3.5Effect of Incubation Time on the Binding

Reagents

All reagents prepared is described in section (2.1.6) and (2.2.3.1).

Procedure

Twenty microliters (100, 75 and 75 μg protein) of cytosolic fraction (fibroadenoma , premenopausal IDC and postmenopausal IDC) were added to 25 μL (0.0565 mg.mL^{-1}) of ^{125}I-anti CA19-9 antibody respectively. The reaction mixture was completed to a final volume 500 μL with PBS buffer pH (7.8 , 8.0 and 7.0) respectively. The experiment was carried out at 25 °C , 37 °C and 45 °C for fibroadenoma , premenopausal (IDC) and postmenopausal (IDC) respectively.

The incubation was carried out at different time intervals (1, 2, 3, 4, 5 and 6 hrs). The bound CA19-9 was estimated as mentioned in section (6.2.2).

Calculations

1.The (B/T) % values were determined as in section (6.2.2).
2.Values of (B/T) % were plotted versus incubation time.

6.2.3.6.Effects of Different Halides on the Binding

Reagents

3. Phosphate buffer (PB) were prepared as described in section (2.1.6) without addition of NaCl .
4. Halid reagents were prepared in concentration of 0.01M PB at pH (7.8, 8.0 and 7.0) individually, by dissolving each of 0.021gm of NaF, 0.0292gm of NaCl, 0.0515gm of NaBr, and 0.075gm of NaI in a final volume 50mL of PB and the pH was adjusted.
5. The breast tumors homogenates (fibroadenoma , premenopausal IDC and postmenopausal IDC) were prepared as described in section (2.1.7), except using PB-

buffer instead of PBS at the same pH and same concentration was carried out the homogenization.

Procedure

The experiment was carried out at optimum conditions as mentioned in section (6.2.3) using three groups of human breast homogenate (i.e., fibroadenoma, premenopausal IDC and postmenopausal IDC), by incubating 20 µL of the homogenate from each group containing (100, 75 and 75 µg protein) respectively with 25 µL (0.0565 mg.mL^{-1}) of ^{125}I-anti CA19-9 antibody. The reaction mixture was completed to a final volume 500 µL with PBS buffer pH (7.8, 8.0 and 7.0) containing 0.01 M of each of the following salts: NaF, NaCl, NaBr and NaI in each assay tube (A sample without the addition of any salt was used as a control). The assay tubes were incubated for (4,1 and 6 h) at 25 ,37 and 45°C for three group individually. The bound CA19-9 was estimated as mentioned in section (6.2.2).

Calculations

1.The (B/T) % values were determined as in section (6.2.2).

2. Values of (B/T) % were plotted versus 0.01 M of NaX.

6.2.3.7. Effects of Monovalent and Divalent Cations on the Binding

Reagents

1. Phosphate buffer (PB) were prepared as described in section (2.1.6) without addition of NaCl .

2. Monovalent and divalent cations (0.025 M) were prepared in PB buffer, and then the pH was adjusted to 7.8, 8.0 and 7.0 individually by dissolving each of 0.0931 gm of KCl, 0.0668 gm of NH_4Cl, 0.2541 gm of $MgCl_2.6H_2O$, 0.1388 gm of $CaCl_2.2H_2O$, 0.2474gm of $MnCl_2.4H_2O$, 0.3150 gm of $CuSO_4.5H_2O$, 0.1703 gm of $ZnCl_2$, in a final volume 50 ml of PB and the pH was adjusted.

Procedure

The experiment was carried out at optimum conditions using three groups of human breast homogenate (i.e., fibroadenoma, premenopausal IDC and postmenopausal IDC) respectively.

The same steps mentioned on section (6.2.3.6) were followed to determine the effect of monovalent and divalent cations on the binding , except ; the buffer solution was PB (0.15 M) containing 0.025 M of the following salts: KCl , NH_4Cl , $MgCl_2.6H_2O$, $CaCl_2.2H_2O$, $MnCl_2.4H_2O$, $CuSO_4.5H_2O$ and $ZnCl_2$.

Calculations

1.The (B/T) % values were determined as in section (6.2.2).

2.Values of (B/T) % were plotted versus the 0.025 M of monovalent and divalent cations.

6.2.3.8 Recovery of CA19-9

Reagents

All reagents prepared is described in section (2.1.6) and (2.2.3.1). Standared concentration of CA19-9 255 $U.mL^{-1}$ was used.

Procedure

The experiment was carried out at optimum conditions. Known concentration of CA19-9 (255 U.mL^{-1}) was added to the three group of benign (fibroadenoma) and malignant (premenopausal IDC and postmenopausal IDC) breast tissues homogenates. The experiment was carried ou at optimum conditions that was obtained in the experiment of section (6.2.3).

Calculations

1. The bound (c.p.m.) of the reaction mixture added to tissue homogenate with ^{125}I-anti CA19-9 antibody, represent the measured value.
2. The bound (c.p.m.) of CA19-9 in tissue homogenate with ^{125}I-anti CA19-9 antibody only , represent the expected value.
3. The recovery % (yield) calculated as follows:

$$Recovery\% = \frac{Measured\,values(c.p.m)}{Expected\,values(c.p.m)}x100$$

Determination of CA19-9 levels in Sera of Patients with Benign and Malignant Breast Tumors

Serum CA19-9 levels were measured by a solid-phase "sandwich" Immunoradiometric Assay (IRMA), which is specifically recognized by the anti CA19-9 monoclonal antibody.The monoclonal antibody is coated on the solid phase, or radiolabeled with the iodine 125 and used as a tracer. The radioactivity of the bound is directly proportional to the amount of CA19-9 presents at the beginning of the assay.

CA19-9 levels in sera of patients with benign breast tumors (group I) and (pre-and post-menopausal) malignant breast tumors (group II and group III) were measured by immunoradiometric assay. Three groups were matched with one group of control subjects. Table (6.2) shows the results obtained from this study. CA19-9 concentration of specimens and control were determined directly from standared curve in figure (6.1) .The level of serum CA19-9 in benign breast tumor patients was found to be 31.0 $U.mL^{-1}$ (p<0.05), where that of (pre-and post-menopausal) malignant breast tumor patients were found to be 33.1 $U.mL^{-1}$ (p<0.05) and 32.1

U.mL^{-1} (p<0.0005) respectively. While in control, the level was found to be 28.8 U.mL^{-1} .Matching case and control subject proved to be important for controlling undesired variability. The mean CA19-9 was significantly high in postmenopausal patients (p<0.0005) while in premenopausal and benign breast tumors the mean of CA19-9 was significantly low (p < 0.05 Student's t-test).

Table (6-2): Sera CA19-9 levels (U.mL^{-1}) in patients with benign and malignant breast tumors. (All other details are explained in the text).

Group	Patients	No. of Cases	Age (year)	Serum CA19-9 U.mL^{-1} (mean ± SD)	P values
I	Benign breast tumors	10	18-35	31.0 ± 1.52	P<0.05
II	Premenopausal malignant breast tumors	10	35-43	33.1 ± 2.79	P<0.05
III	Postmenopausal malignant breast tumors	10	53-65	32.1 ± 0.13	P<0.0005
Control	Control	10	25-35	28.8 ± 0.631	

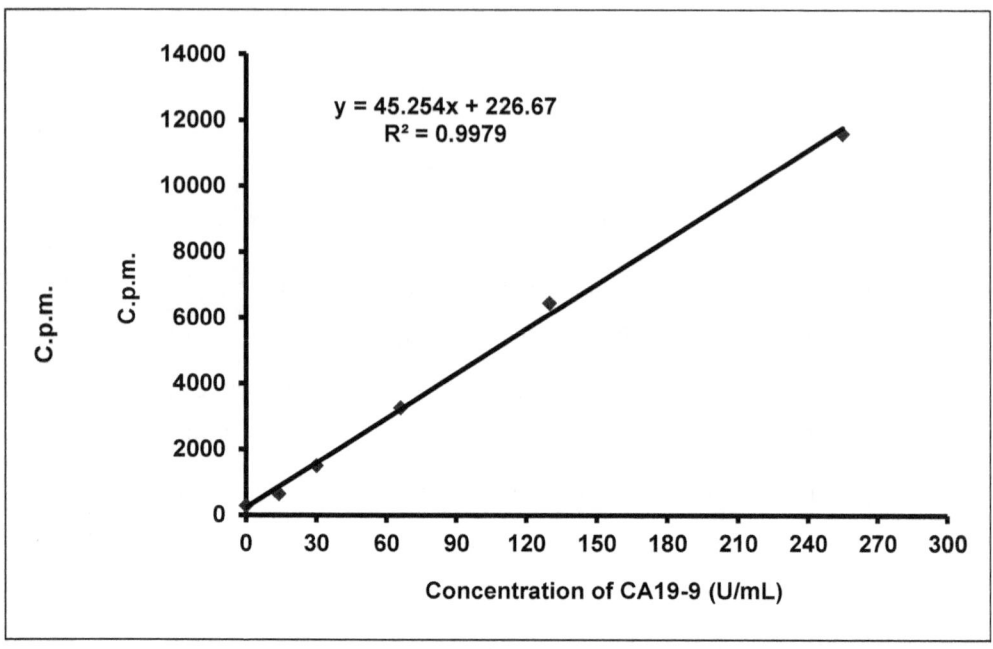

Figure (6.1): Standard curve of CA19-9. (All other details are explained in the text).

CA19-9 was at low concentration in sera of healthy individuals, these results are in agreement with several authors previously [120].

There were few studies to evaluate CA19-9 in breast tumors patients. Several investigators [222] detected CA19-9 in bone metastasis in breast cancer patients and in patients without documented metastases and reported that CA19-9 level elevated in patients with metastases breast cancer.

When patients were analyzed with respect to the menopausal status, significant differences between the monastic and non monastic patients was detected [222].

Several studies proved the possibility of the role of carbohydrate antigen 19-9 as a tumor marker in colorectal cancer [132], pancreas [234], gastric [235], liver disease [236] and esophageal cancer [125]. Recently, European group proved that CA19-9 monitored in patients with tumors of gastrointestinal tract and endometrial cancer could be used as a tumor marker and can be helpful [237] in monitoring patients with breast cancer. They observed significant increase of CA19-9 and CA15-3 levels [237] in all patients.

Preliminary Test of the Binding of CA19-9 with 125I-Anti CA19-9 Antibody

Supernatant and pellet obtained at speed (4000 r.p.m) were investigated in the three groups of human breast tumor homogenate (fibroadenoma, premenopausal IDC and postmenopausal IDC). In each fraction, CA19-9 was detected through the incubation of ^{125}I-anti CA19-9 antibody with crude fraction supernatant and pellet individually for 3 h at 37°C in PBS buffer pH 7.2 as a medium to complete the reaction.

The separation of the bound antibody from unbound was carried out at 4000 r.p.m for 45 min. to precipitate the ^{125}I-anti CA19-9 antibody/CA19-9 complex formed.

Table (6.3): Incidence of CA19-9 in supernatant and pellet fractions in three different breast homogenate. (All other details are explain in the text) .

Groups	Age(year)	B/T %	
		Supernatant fraction	Pellet fraction
Benign	34	5.32	1.43
Premenopausal (IDC)	43	5.48	2.03
Postmenopausal (IDC)	63	5.86	2.47

Table (6-3) shows the amount of binding B/T % values of pellet and supernatant fractions. The data revealed that CA19-9 in cytosolic fraction obtained from supernatant was higher in incidence than in pellet fraction, according to these results cytosolic fraction was collected. CA19-9 collected and the pellet was then discarded.

Factors Effecting of 125I-Anti CA19-9 Antibody Binding to CA19-9 in Breast Tumors Homogenates

Effect of Protein Concentration on the Binding

To obtain the optimum protein concentration of cytosolic fraction for the binding of CA19-9 with ^{125}I-anti CA19-9 antibody, cytosolic fraction containing increasing amount of soluble CA19-9 in the presence of fixed amount of ^{125}I-anti CA19-9 antibody was carried out as it was mentioned in section (6.2.3.1). Figure (6-2) represent the formation of (^{125}I-anti CA19-9 antibody/CA19-9) complex in three cases (fibroadenoma, premenopausal IDC and postmenopausal IDC) and shows that (100, 75 and 75 µg protein) were the most appropriate concentration to give the maximum values of binding in crude fraction of three cases respectively. The decrease of the binding at high concentration of cytosolic fraction (in three cases) in the reaction mixture may be due to a conformational change in CA19-9 and ^{125}I-anti CA19-9 antibody rather than the formation of reversible inactive (^{125}I-anti CA19-9 antibody/CA19-9) complex [238] and may be due to splitting antigen into large fragments with proteolytic enzymes [239].

In all subsequent experiments an amount of (100, 75 and 75 µg protein in three cases respectively), were used in the incubation mixture.

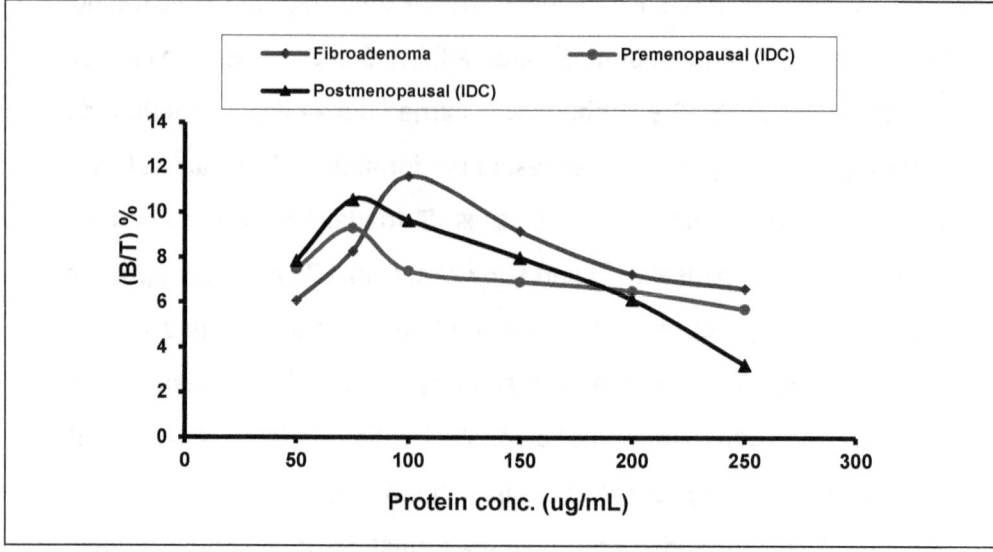

Figure (6.2): Influence of increasing protein concentrations on the binding of CA19-9 with ^{125}I-anti CA19-9 antibody. (All other details are explained in the text).

Effect of 125I-Anti CA19-9 Antibody concentration on the Binding

One of the most important factors that effect binding is the concentration of ^{125}I-anti CA19-9 antibody. To determine the suitable concentration of ^{125}I-anti CA19-9 antibody, cytosolic sample (100, 75 and 75 µg protein) in the three cases (fibroadenoma, premenopausal IDC and postmenopausal IDC) respectively were incubated with increasing concentration of ^{125}I-anti CA19-9 antibody, the incubation was carried out for 3 h at 37 °C. The results revealed that the optimum concentration of the ^{125}I-anti CA19-9 antibody to give the maximum binding in all three cases was (0.0565 mg.mL^{-1}). The results showed that an increase in the conc. of ^{125}I-anti CA19-9 antibody caused a decrease in the binding %. This is because the soluble complexes, and the excess of antibody cover all antigentic sites, which leads to complex formation inhibition. Accordingly in all subsequent experiments, 0.0565 mg.mL^{-1} of ^{125}I-anti CA19-9 was used as the optimum conc., which gives the highest binding %.

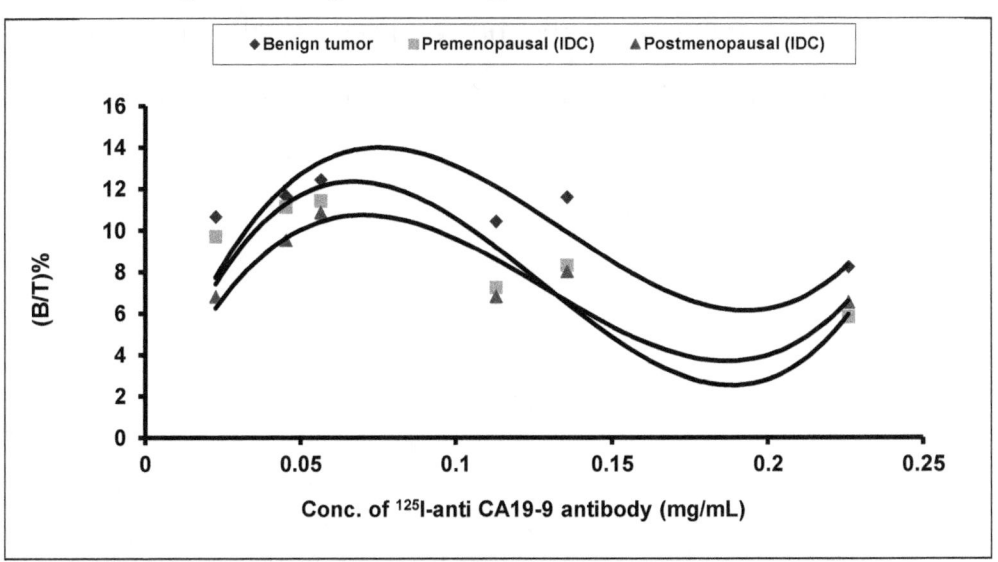

Figure (6.3): Effect of different concentration of ^{125}I-anti CA19-9 antibody on the binding with CA19-9. (All other details are explained in the text).

Effect of pH on the Binding

The effect of pH on the binding of radioactivity CA19-9 to its antigen CA19-9 was investigated. Figure (6-4) shows that the maximum binding of ^{125}I-anti CA19-9 antibody to its antigen CA19-9 was found to be (7.8, 8.0 and 7.0) in the three cases used (fibroadenoma , premenopausal IDC and postmenopausal IDC) respectively. The shift in pH of the environment may involve a protonation-deprotonation process occuring within the change of polar groups of the amino acids residues present in the binding domain [240] .According to these results, the pH of the buffer used in all subsequent experiments were (7.8, 8.0 and 7.0) for the three cases respectively.

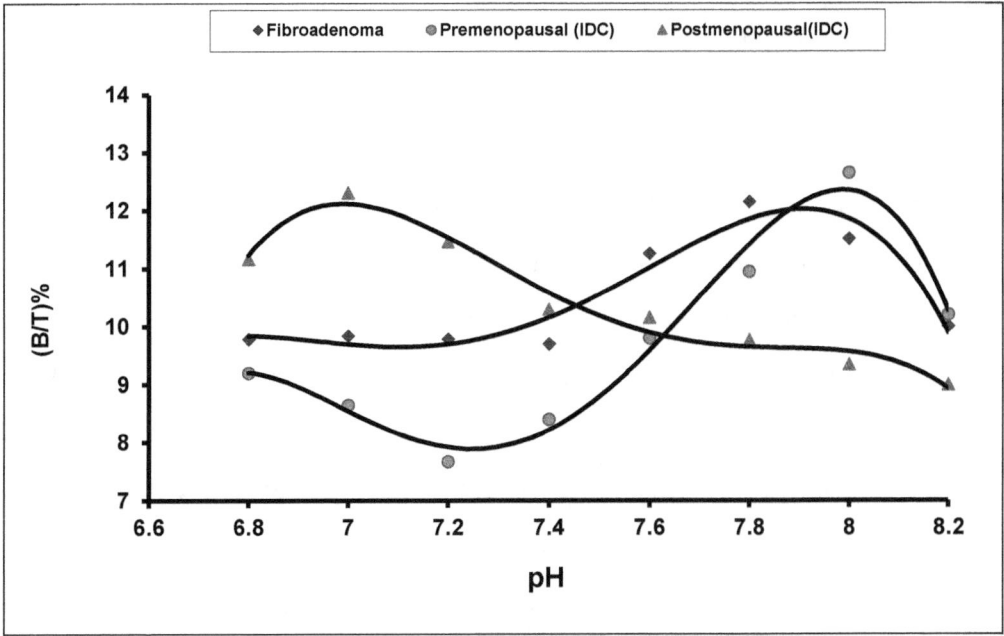

Figure (6-4): The effect of pH on the binding of CA19-9 with its antibody ^{125}I-anti CA19-9 antibody with CA19-9. (All other details are explained in the text).

Effect of Temperature on the Binding

Temperature dependency of the association of ^{125}I-anti CA19-9 antibody to its cytosolic fraction CA19-9 was investigated. Cytosol fraction of benign and malignant breast tumors was incubated for 3 hrs at different temperatures (5, 15, 25, 37 and 45 °C). Figure (6-5) reveals that the binding of ^{125}I-anti CA19-9 antibody to its cytosol fraction CA19-9 was increased when the temperature was raised from 5 to 25 °C in fibroadenoma and the maximal binding was obtained at 25 °C and from 5 to 37 °C in premenopausal (IDC) and the maximal binding was obtained at 37 °C. Finally from 5 to 45 °C in postmenopausal (IDC) and the maximal binding was obtained at 45 °C. The decrease in the binding at temperature higher than the optimum temperature is probably due to denaturation of CA19-9 molecules [241] or due to proteolytic degradation of enzyme [150]. According to these results (25 °C , 37 °C and 45 °C) respectively they will be used in all the subsequent experiments for the three cases used.

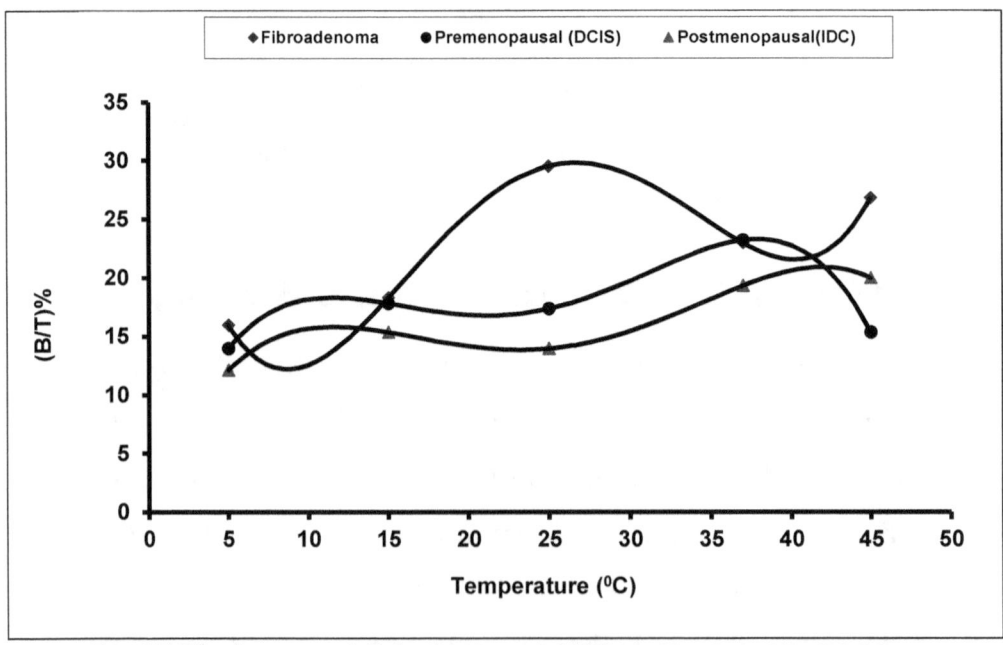

Figure (6-5): Effect of temperature on the binding of **125I-anti** [CA19-9] **antibody with CA19-9**. (All other details are explained in the text).

The Effect of Incubation Time on the Binding

To choose the most appropriate incubation time at (25, 37 and 45 °C) for the three cases used in this study (fibroadenoma, premenopausal IDC and postmenopausal IDC) respectively , the experiments were carried out at different time intervals. Figure (6-6) shows the results of this analysis. It seemed that the specific binding of ^{125}I-anti CA19-9 antibody to cytosolic fraction homogenate for the three cases were maximal at (4,1 and 6 hrs) respectively. In view of these results, the incubation time used in all subsequent experiments were (4,1 and 6 hrs) respectively.

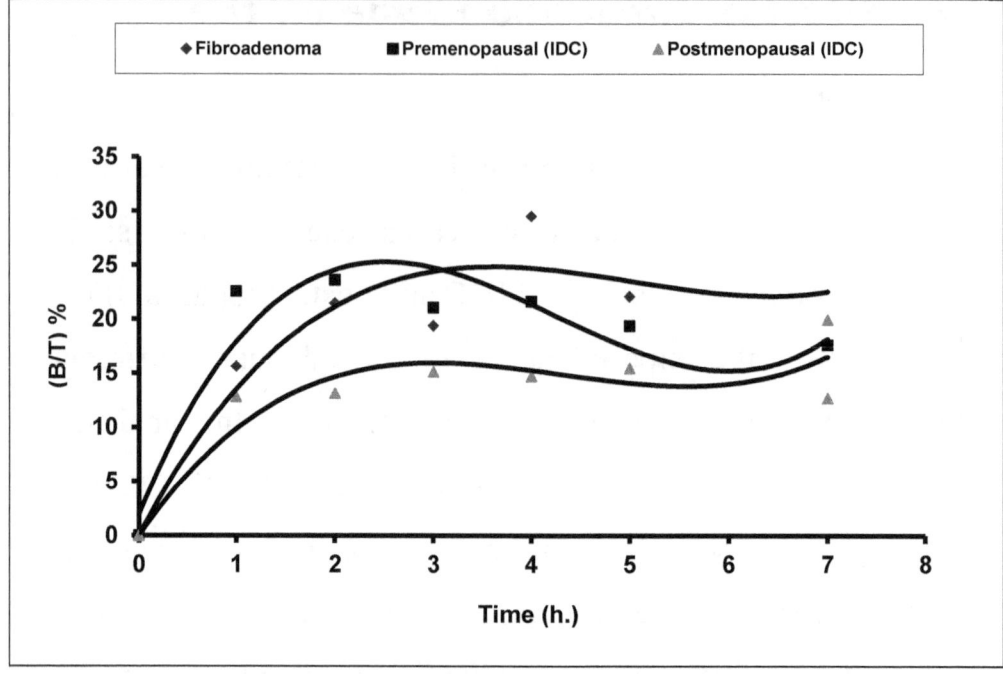

Figure (6.6): The effect of incubation time on the binding of **^{125}I-anti** $^{CA19-9}$ **antibody with CA19-9. (All other details are explained in the text).**

Effect of Different Halides on the Binding

Figure (6.7) shows the effect of different halides salts (i.e., NaF, NaCl, NaBr and NaI) at 0.01 M concentration on the extent of ^{125}I-anti CA19-9 antibody binding to their cytosol fraction homogenate in benign and malignant breast tumors. The sodium halides (ion radius) in the incubation mixture of benign and postmenopausal malignant breast tumors induced inhibition of the percent of binding according to the following sequence:

NaI>NaF>NaCl>NaBr

While the sodium halides in the incubation mixture of premenopausal malignant breast tumor (IDC) induced activation of the percent of the binding in the order:

NaF<NaCl<NaI<NaBr

Melander and Horvath (1977) reported that the effect of halide salt type on hydrophabic interactions is quantified by its molar surface tension increment (MSTI) which is a measure of the increasing in a surface tension by the salt [171], also they found that parameter increases as the following sequence:

NaF>NaCl>NaI

The same researches found that halides with higher MSTI values will strengthen the hydrophabic interactions while halides with lower MSTI values reverse this effect. Thus the dependence of the extent of the binding in benign and malignant (pre-and post-menopausal) breast tumors on MSTI values of the corresponding halide further implicates the low involvement of hydrphobic forces in maintaining the stability of (^{125}I-anti CA19-9 antibody /CA19-9) complex formed.

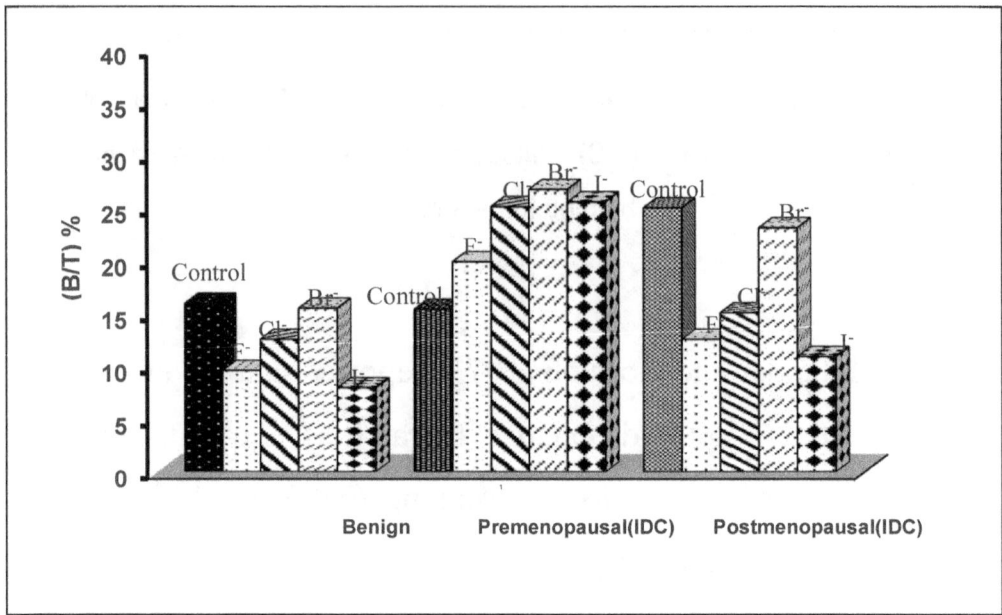

Figure (6-7): Effect of different halides on the binding of of
125I-anti CA19-9 **antibody with CA19-9. (All other details are explained in the text).**

Effect of Monovalent and Divalent Cations on the binding

Figure (6.8) and (6.9) show the effect of different divalent and monovalent cations respectively on the binding value in benign and malignant breast tumors. The results indicate that the binding process is sensitive to the presence of cation metal ions. $CuSO_4.5H_2O$ at concentration 25 mM was showed to increase the binding two folds than the control as comparied with other divalent cations.

$CaCl_2.2H_2O$ induced activation in the binding in benign (fibroadenoma) and malignant (premenopausal IDC), while induced inhibition in the binding in malignant (postmenopausal IDC). $ZnCl_2$ decreased the binding in two groups (fibroadenoma and premenopausal IDC) , while $ZnCl_2$ increased the binding in malignant (postmenopausal IDC).

The frequency of the stimualtion of the binding of ^{125}I-anti CA19-9 antibody to its cytosolic fraction CA19-9 homogenate of the three groups by divalent cations is according to the following:

Postmenopausal breast cancer tissue homogenate (IDC)
$$Cu^{+2}>Zn^{+2}>Mn^{+2}>Mg^{+2}>Ca^{+2}$$
Premenopausal breast cancer tissue homogenate (IDC)
$$Cu^{+2}>Ca^{+2}>Mn^{+2}>Mg^{+2}>Zn^{+2}$$
Benign breast tumor tissue homogenate (Fibroadenoma)
$$Cu+2>Ca+2>Mg+2>Mn+2>Zn+2$$

The binding of metal ions to proteins is a function of pH among the different classes of groups, such as carboxyl, amino, imidozol and tyrosyl (the unshared electron pairs for nitrogen , oxygen and sulfur atoms) [242]. The sites of binding of metal ions may range from elaborate chelate sites to simple complex formation which discrete single ligand groups in the protein. In short, chelation plays a dominant role in establishing the relative strengths of binding of a given metal ion by various sites in protein [243].

Figure (6-9) shows that monovalent cations inhibite the binding in benign and malignant premenopausal (IDC), while the monovalent cations induce activation of the binding in-group of malignant postmenopausal (IDC). The alternation of increased and decreased binding percent between these cations may be ascribed to the differences in tissues studied [244]. The variation of results obtained between these divalent cations may be ascribed to the difference in tissue studied [245]

.

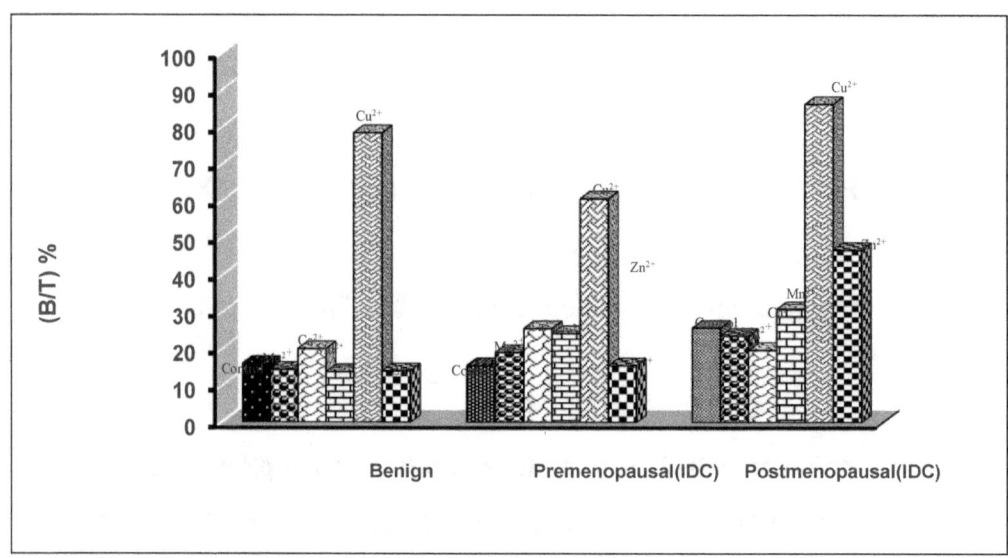

Figure (6.8): Effect of different cations on the binding of **^{125}I-anti** [CA19-9] **antibody with CA19-9 in different human breast tumor homogenate. (All other details are explained in the text).**

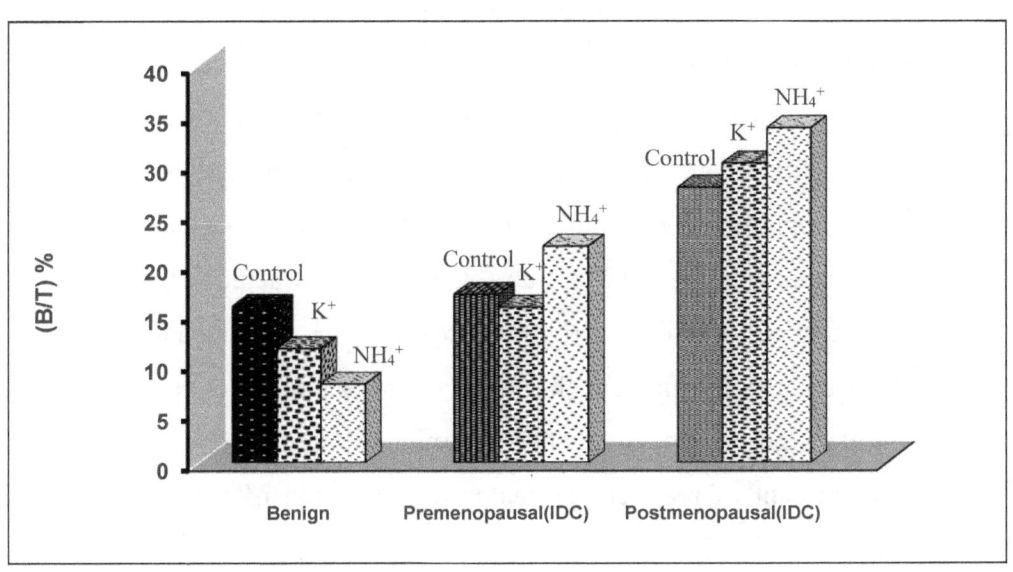

Figure (6.9): Effect of different monovaleat cations on the binding of 125**I-anti** CA19-9 **antibody with CA19-9 in different human breast tumor homogenate. (All other details are explained in the text).**

Recovery of CA19-9

The method used to estimate the percent recovery of cytosolic fractions of benign (Fibroadenoma) and malignant (pre-and post-menopausal IDC) breast tumors homogenates. The results summarized in table (6.4) indicate that CA19-9 extracted from malignant breast tumors homogenates was recovered less than CA19-9 extracted from benign breast tumor (fibroadenoma) and CA19-9 extracted from malignant premenopausal (IDC) malignant breast tumors homogenates was recovered more than CA19-9 extracted from postmenopausal (IDC) malignant breast tumors homogenates. Also the results indicate that total CA19-9 could determine through the developed method of immunoradiometric assay. The percent of recovery indicates the precision of the used method.

Table (6.4): Recovery of CA19-9 (All details are explained in the text).

Type of CA19-9	Measured B/T %	Expected B/T %	Recovery % Measured/ Expected %
Benign	103	104	99
Premenopausal	192	195	98
Postmenopausal	166	175	95

Conclusion

1. The developed protocol for the assay of CA15-3 or CA19-9 is suitable for the assessment of CA15-3 or CA19-9 in tissues.

2. The magnitude of elevation in CA15-3 is more than CA19-9 in the same sera of patients with breast tumors. This indicates that CA15-3 as a tumor marker is more specific than CA19-9.

3. Partially purified CA15-3 from benign and malignant breast tumors homogenates shows high affinity to ^{125}I-anti CA15-3 antibody than crude CA15-3.

4. Kinetic studies of ^{125}I-anti CA15-3 antibody with partially purified CA15-3 in benign and malignant breast tissues homogenates show that the reaction is temperature and time dependent. Pseudo first order kinetics at (5, 15, 25, 37, and 45 °C) was observed, in all cases.

5. The results obtained from the thermodynamic studies on the association of ^{125}I-anti CA15-3 antibody with partially purified CA15-3 in benign and malignant breast tissues homogenates indicate that the binding reaction occurs spontaneously $\Delta G° < 0$, and entropically driven since $\Delta S° > 0$.

6. The spectroscopic studies revealed that the (^{125}I-anti CA15-3antibody/CA15-3) complex in the two cases benign and malignant breast tissues have a characteristics spectrum.